Domain Oriented Systems Development

Advanced Information Processing Technology
A series edited by Tadao Saito

Volume 1

Domain Oriented Systems Development – principles and approaches
Edited by K. Itoh, T. Hirota, S. Kumagai and H. Yoshida (1998)

Volume 2

Designing Communications and Collaboration Support Systems
Edited by Y. Matsushita (1999)

Volume 3

Information Networking in Asia
Edited by H. Higaki, Y. Shibata and M. Takizawa (2001)

Volume 4

Advanced Lisp Technology
Edited by T. Yuasa and H. G. Okuno (2002)

Volume 5

Nontraditional Database systems
Edited by Y. Kambayashi, M. Kitsuregawa, A. Makinouchi, S. Uemura, K. Tanaka and Y. Masunaga (2002)

Also available from Taylor & Francis

Professional Issues in Software Engineering 3rd edition
by Frank Bott et al. (2000)

Deductive Databases and their Applications
by Robert Colomb (1998)

Theory and Practice of Relational Databases
by Stefan Stanczyk, Bob Champio and Richard Leyton (2001)

Introduction to Artificial Intelligence
by J. Finlay and A. Dix (1996)

Domain Oriented Systems Development

Practices and perspectives

Edited by

Kiyoshi Itoh, Satoshi Kumagai, and Toyohiko Hirota

CRC Press
Taylor & Francis Group
Boca Raton London New York

CRC Press is an imprint of the
Taylor & Francis Group, an **informa** business
A TAYLOR & FRANCIS BOOK

First published 2003 by Taylor & Francis

CRC Press
Taylor & Francis Group
6000 Broken Sound Parkway NW, Suite 300
Boca Raton, FL 3487-2742

ISBN-13: 978-0-415-30450-4 (hbk)

This book has been produced from camera-ready-copy
supplied by the authors

Visit the Taylor & Francis Web site at
http://www.taylorandfrancis.com

and the CRC Press Web site at
http://www.crcpress.com

British Library Cataloguing in Publication Data
A catalogue record for this book is available from the British Library

Library of Congress Cataloging in Publication Data
A catalog record for this book has been requested

Contents

Foreword to the series

The Information Processing Society of Japan (IPSJ) is the top academic institution in the information processing field in Japan. It has about thirty thousand members and promotes a variety of research and development activities covering all aspects of information processing and computer science.

One of the major activities of the society is publication of its transactions containing papers covering all the fields of information processing, including fundamentals, software, hardware, and applications. Some of the papers are published in English, but because the majority are in Japanese, the transactions are not suitable for non-Japanese wishing to access advanced information technology in Japan.
IPSJ therefore decided to publish a book series titled "Advanced Information Processing Technology."

The series consists of independent books, each including a collection of top quality papers, from mainly Japanese sources of a selected area in information technology. The book titles were chosen by the International Publication Committee of IPSJ so that they enable easy access to information technology for international readers. Each book contains original papers and/or those updated or translated from original papers appearing in the transactions or internationally qualified meetings. Survey papers to aid understanding of the technology in Japan in a particular area are also included.

As the chairman of the International Publication Committee of IPSJ, I sincerely hope that the books in the series will improve communication between Japanese and non-Japanese specialists for their mutual benefit.

Tadao Saito

Chairman
International Publication Committee
The Information Processing Society of Japan

Preface

This book "Domain Oriented Systems Development - Practices and Perspectives" is the second collection of articles edited by the research group on domain analysis and modeling in IPSJ SigSE (Information Processing Society of Japan, Special Interest Group of Software Engineering).

This research group has been studying the concept and applicability of domain oriented systems development since 1993 until 2001. The research group published a book in Japanese language in 1996 summer, and published a book in English language in 1998 autumn that is the first collection of articles by the research group, "Domain Oriented Systems Development - Principles and Approaches". The research group organized DOSD 1999 International Workshop in Asia-Pacific Software Engineering Conference by IPSJ SigSE at Takamatsu, Japan in November 1999.

This book mainly includes the extended versions of several presentations by International Workshop participants from Japan, Korea, Canada, and Austria and the qualified papers selected by the research group. This book covers the wide range of activities on domain oriented systems development, i.e., triplex schema of key concepts for domain oriented systems development, feature oriented product line software engineering, software architecture for intelligent CAD, fully independent component architecture, interrelationships between requirements engineering and domain engineering, framework development and evolution, knowledge base development for product configuration systems, domain specific repository and reuse for CAI, domain framework based on three-view model, business process analysis using IDEF0, generation of architecture product model.

The editors wish to thank IPSJ SigSE, former chairperson Prof. Tetsuo Tamai (University of Tokyo) and present chairperson Prof. Mikio Aoyama (Nanzan University), for supporting the research group on domain analysis and modeling.

Kiyoshi Itoh, Satoshi Kumagai, Toyohiko Hirota

Contributors

Editors

Kiyoshi Itoh, a co-editor of this volume, is a professor at Sophia University, Tokyo, Japan. He received B. Eng., M. Eng., Dr. Eng. in computer science all from Kyoto University in 1974, 1976 and 1980, respectively. His research interests include software engineering, software prototyping, domain analysis and engineering, collaboration and concurrent engineering, simulation methodology, expert system, etc. He is a member of IPSJ, IEEE, ACM, SDPS. He can be reached at itohkiyo@me.sophia.ac.jp.

Satoshi Kumagai, a co-editor of this volume, is a researcher at Research and Development Headquarters of Yamatake Corporation. He received his B.E and M.E in Administration Engineering from Keio University in 1984 and 1986, respectively. He earned a doctoral degree from Sophia University in 2000. His research interest includes domain analysis and modeling, software engineering, and collaboration and concurrent engineering. He is a member of IEEE, SDPS, etc. He can be reached at satoshi_kumagai@nifty.com.

Toyohiko Hirota, a co-editor of this volume, is a professor at Kyushu Sangyo University, Fukuoka, Japan. He received his B.E. in electrical engineering in 1976 and his M.E. and Dr. Eng. in computer science in 1978 and 1984, respectively, all from Kyoto University. His research interests include software engineering, domain analysis and modeling, software testing, database schema design, etc. He is a member of IPSJ, IEICE, IEEE CS, etc. He can be reached at hirota@is.kyusan-u.ac.jp.

Chapter 1

Ryo Kawabata is a research associate at Sophia University, Tokyo, Japan. He received his B.E. and M.E. in mechanical engineering from Sophia University, in 1996 and 1998, respectively. His research interests include software engineering, domain analysis and modeling, collaboration and concurrent engineering. He is a member of IPSJ, SCS. He can be reached at r-kawaba@me.sophia.ac.jp.

Akiko Hasegawa is a doctor-course student at Sophia University. She received a B.E. and M.E. in Mechanical Engineering from Sophia University, Japan, in 1998 and 2000, respectively. Her research interests are domain analysis and modeling, and concurrent engineering. She can be reached at a-hasaga@me.sophia.ac.jp.

Manabu Kamimura is a graduate student at Sophia University. He received a BS in Mechanical Engineering from Sophia University, Japan, in 2000. Research interests are domain analysis and modeling, and concurrent engineering.

Chapter 2

Kyo C. Kang is a professor at Pohang University of Science and Technology, Pohang, Korea. He received his B.S. in Statistics in 1973 from Korea University, M.S. in engineering design in 1975 from University of Colorado, and Ph.D. in industrial engineering in 1982 from the University of Michigan. He was a Member of Technical Staff at Bell Communications Research (Bellcore) and Bell Labs, AT&T, and a Senior Member of Technical Staff at the Software Engineering Institute (SEI), Carnegie Mellon University. His area of research includes CASE, requirements engineering, software reuse, domain analysis, and product line engineering. His is a member of the Korea Information Science Society, ACM, and IEEE.

Kwanwoo Lee is a PhD. Candidate at Pohang University of Science and Technology (POSTECH), Pohang, Korea. He received his B.S. M.S. in computer science and engineering from POSTECH in 1994 and 1996, respectively. His research interests include software reuse, domain engineering, product line engineering, architecture design and analysis, real-time systems development, etc. He can be reached at kwlee@postech.ac.kr.

JaeJoon Lee is a Ph.D. candidate at Pohang University of Science and Technology (POSTECH), Pohang, Korea. He received his B.S. in mathematics from Sogang University in 1991 and M.E. in computer and communications engineering from POSTECH in 1998. He was an associate engineer at LG Information and Communications Ltd., R&D center from 1993 to 2000. His research interests include software reuse, software product line engineering, domain engineering, software architecture, and configuration management. He can be reached at gibman@postech.ac.kr.

Sajoong Kim is a PhD. candidate at Pohang University of Science and Technology (POSTECH), Pohang and a researcher at KIPA (Korea IT Industry Promotion Agency). Seoul, Korea. He received his B.E and M.E in computer science from Kwang-Woon University and Chung-Ang University, in 1984 and 1986, respectively. His research interests include software engineering, domain analysis and feature modeling, software architecture design, etc. He can be reached at sjkim@software.or.kr.

Chapter 3

Masa-aki Hashimoto is a professor at Kyushu Institute of Technology. He received B. Eng., M. Eng., Dr. Eng. in computer engineering all from Kyushu University in 1968, 1970 and 1987 respectively. He worked for Nippon Telegraph and Telephone Corporation from 1970 to 1992. His research interests include software engineering, business process analysis and reengineering, project management and so on. He is a member of IEEE, IPSJ, IEICE, JSSST and JSAI. He can be reached at hasimoto@ai.kyutech.ac.jp.

Chapter 4

Yoshiaki Fukazawa was born in Shizuoka, Japan, on July 17, 1953. He received the B.E., M.E., and D.E. degrees in electrical engineering in 1976, 1978, and 1986, respectively all from Waseda University. He is currently a professor in School of Engineering and Science, Waseda University. His major interests are software engineering, fine-grain parallel architecture and its optimizing compiler, and software support of hardware design. He is a member of IPSJ, IEICE, IEEE, ACM and so on. He can be reached at fukazawa@fuka.info.waseda.ac.jp.

Hironori Washizaki was born in Tokyo, Japan, on November 19, 1976. He received the Bachelor's and Master's degrees in information and computer science in 1999 and 2001, respectively, all from Waseda University, Tokyo, Japan. He is currently a doctor course's student in School of Engineering and Science, Waseda University. His major interest is software engineering, especially component-based software development and software patterns. He is a member of IEICE and JSSST. He can be reached at washi@fuka.info.waseda.ac.jp.

Chapter 5

Atsushi Ohnishi received B.E., M.E. and Dr. Eng. degrees from Kyoto University in 1979, 1981, and 1988, respectively. He was a research associate of Kyoto University from 1983 to 1989 and an associate professor of Kyoto University from 1989 to 1994. Since 1994 he has been a professor in Department of Computer Science at Ritsumeikan University. His current research interests include requirements engineering, object-oriented analysis, and software design techniques. He is a member of IEEE computer society, ACM, IPSJ, IEICE, and JSSST. His e-mail address is ohnishi@cs.ritsumei.ac.jp.

Chapter 6

Gregory Butler is Professor in Computer Science at Concordia University, Montreal, Canada. He joined Concordia in 1992 after nine years on the faculty of the University of Sydney, Australia. He received a B.Sc(Hons) in 1976 and a Ph.D. in 1980 from the University of Sydney. His research interests include software technology for complex software systems, scientific knowledge based systems, database technology, and bioinformatics. He is a member of International Society for Computational Biology. He can be reached at gregb@cs.concordia.ca.

Ling Chen is a software designer. She received her B.Eng in electrical engineering from Shenyang Polytechnic University in 1984, and Master of Computer Science from Concordia University in 2001. Her research interests include software engineering and database systems. She can be reached at Ling_Chen8@hotmail.com.

Xuede Chen is a senior engineer at Nortel Networks, Ottawa, Canada. He received a B.Eng. in computer science from Shenyang Polytechnic University in 1984, and a M.Eng. in computer science from Northeastern University in 1988. His research interests include database systems, user interface design, system capacity and performance engineering. He can be reached at xuede@nortelnetworks.com.

Ashraf Gaffar is a part-time instructor at Concordia University, Montreal, Canada. He received a B.Sc. in electrical engineering from Cairo University, Egypt in 1988 and a M.Comp.Sc. from Concordia University in 2001. His research interests include software engineering and databases. He can be reached at gaffar@vif.com.

Jinmiao Li is a software designer at Alcatel Canada. She received her B.Sc. from Fundan University, China in 1992 and M.Comp.Sc. from Concordia University, Canada in 2001. She is now pursuing her Ph.D. study at Concordia University. Her research interests include object-oriented software design and reusable software architecture. She can be reached at li_j@cs.concordia.ca.

Lugang Xu is a Ph.D. candidate at Concordia University, Montreal, Canada. He received a Bachelor degree in software in 1994 and a Master degree in software engineering in 1997, respectively, from Jilin University, Changchun, China. His research interests include software engineering, framework development, domain analysis, and database technology. He can be reached at lxu@cs.concordia

Chapter 7

Gerhard Friedrich is professor of computer science at the University Klagenfurt, Austria. He directs the Computer Science and Manufacturing research group. From 1991 to 1993 he was a group leader of the Advanced Reasoning Methods group of the Christian Doppler Laboratory for Expert Systems at the Vienna University of Technology. From 1993 to 1997 he managed the Department for Configuration and Diagnosis Systems at Siemens Austria. Gerhard Friedrich led numerous industrial R&D-projects and was a research fellow at Siemens corporate research and at the Stanford Research Institute. He received a PhD and a MS in computer science from Vienna University of Technology, Austria. His research interests include configuration, diagnosis and repair, knowledge acquisition and distributed artificial intelligence. Gerhard Friedrich can be reached at gerhard.friedrich@uni-klu.ac.at.

Alexander Felfernig is research assistant of computer science at the University Klagenfurt , Austria. He received a PhD in 2001 and a MS in computer science from the University Klagenfurt in 1997. In his PhD thesis he developed approaches for integrating software engineering and AI techniques in knowledge-based configuration. His research interests include configuration, knowledge acquisition, distributed AI, and ontologies. He is a member of IEEE. Alexander Felfernig can be reached at alexander.felfernig@uni-klu.ac.at.

Dietmar Jannach is research assistant of computer science at the University Klagenfurt, Austria. He received a PhD in 2001 and a MS in computer science from the University Klagenfurt in 1998. In his PhD thesis he developed approaches for supporting debugging of configuration knowledge bases using model-based diagnosis techniques. His research interests include configuration, model-based diagnosis, and distributed configuration. Dietmar Jannach can be reached at dietmar.jannach@uni-klu.ac.at.

Markus Zanker is research assistant of computer science at the University Klagenfurt, Austria. He received a MS in computer science in 1998 and a MS in business administration in 2001. His research interests include distributed configuration, agent-based systems, and ontologies. Markus Zanker can be reached at markus.zanker@uni-klu.ac.at.

Chapter 8

Takashi Fuji, is a professor at Hokkaido Information University, Hokkaido, Japan. He received his B.S. in mathematics in 1969 from Sophia University. He earned a doctoral degree from Sophia University in 1998. His research interests include software engineering, domain analysis and modeling, e-education methodology, etc. He is a member of IPSJ, IEICE, JSAI. ACM, IEEE CS, etc. He can be reached at fuji@do-johodai.ac.jp.

Chapter 9

Takeo Hayase is a researcher at Systems Integration Technology Center of e-Solutions Company, Toshiba Corporation. He received his B.E. and M.E. degrees in the Department of Mechanical Engineering from Sophia University in 1990 and 1992, respectively. He earned a doctoral degree from Sophia University in 2002. His research interests include object-oriented technology, domain analysis, domain engineering, and software engineering. He is a member of IPSJ, IEEJ, IEEE, ACM, etc. He can be reached at takeo.hayase@toshiba.co.jp

Chapter 11

Takeshi Ito received his B.E. and M.E. degrees in computer science from Kyushu Institute of Technology in 1999 and 2001, respectively. He is currently a student in Doctor's program at Kyushu Institute of technology. His research interests are computer aided design, object modeling, and knowledge representation. He can be reached at itoh@minnie.ai.kyutech.ac.jp

CHAPTER 1

An Approach for Application of Domain Oriented Systems Development Using Triplex Schema of Key Concepts

Kiyoshi Itoh

Ryo Kawabata

Akiko Hasegwa

Manabu Kamimura

Sophia University

Satoshi Kumagai

Yamatake Corporation

ABSTRACT

An approach for domain oriented systems development is shown by a triplex schema of concept - D-AME framework, Triadic Domain Model, and Generic Task. These concepts can be considered orthogonal and enable to organize diversified practices of domain oriented systems development. D-AME Framework defines the processes and deliverables in order to enforce the management aspects of domain analysis and modeling activities. Triadic Domain Model specifies the structure of domain model in terms of problem, process, and product of the domain. Generic Task clarifies reusable essentials of domain to facilitate recycling of domain model across multiple application domains.

The authors have been applying the triplex scheme of key concepts to various domains. In the process of these applications, the authors have developed many tools that exemplify domain development facilities. Each of these tools has one or more generic tasks to be analyzed and designed, one or more charts to be used for analyzing and designing systems in domains, and one or more process models to be used.

1.1 INTRODUCTION

The cost and time for systems development can be often reduced by reusing and modifying specifications or source codes of the similar systems in past practices for systems development. Currently, the spectrum of application domains by systems development practices is widely scoped. It is important for systems development to obtain the characteristics, properties, constraints and systems structures specific to each application domain. Domain analysis and modeling is a method which aims to develop systems efficiently by analyzing and modeling these characteristics, properties and systems structures specific to each application domain. Domain analysis and modeling can construct and reuse the models for the requirements analysis and specification, the design specification and diagram, the programming process and source code, the algorithm, DB schema, the parameter tuning process, the development process method, etc.

Domain analysis and modeling ((Prieto-Diaz, 1987, 1991) (Tracz, 1992) (Itoh, 1995, 1998), etc.) is the process which obtains the domain specific model with the terminology, the problem solving / system analysis / system design strategies, the target system structures, etc. It is obtained by analyzing, identifying and organizing the properties of a family of target systems and the heuristic and reasonable knowledge for a family of past practices of systems development. Its purpose is to reduce efforts and time for systems development and to improve the software productivity and reusability effectively.

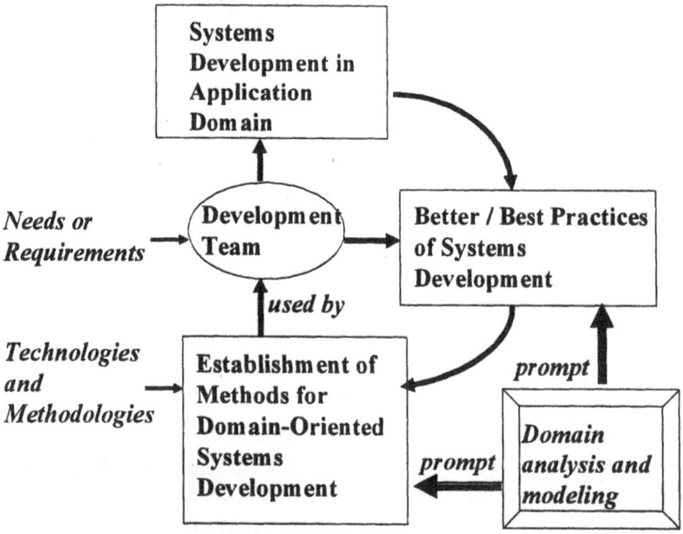

Figure 1.1 Cycle for Domain Oriented Systems Development (DOSD Cycle)

Figure 1.1 shows the cycle for Domain Oriented Systems Development (DOSD Cycle). This name of DOSD Cycle is devised by the authors, but the concept of DOSD cycle has been used, modified and cultivated by the wide range of software engineering practitioners. In Figure 1.1, development teams construct systems in their application domains. They gather and organize best or better practices of systems development. They devise or establish domain oriented methods for systems development. Then they can use such methods for constructing systems in similar application domains. DOSD Cycle shows the reuse activity by using appropriate methods for domain oriented systems development which accumulate and organize best or better practices of systems development.

The size or range of one domain should be carefully determined. The estimation for the number of prospective systems to be developed in the domain is very important. The adoption of DOSD cycle is better where the number is enough.

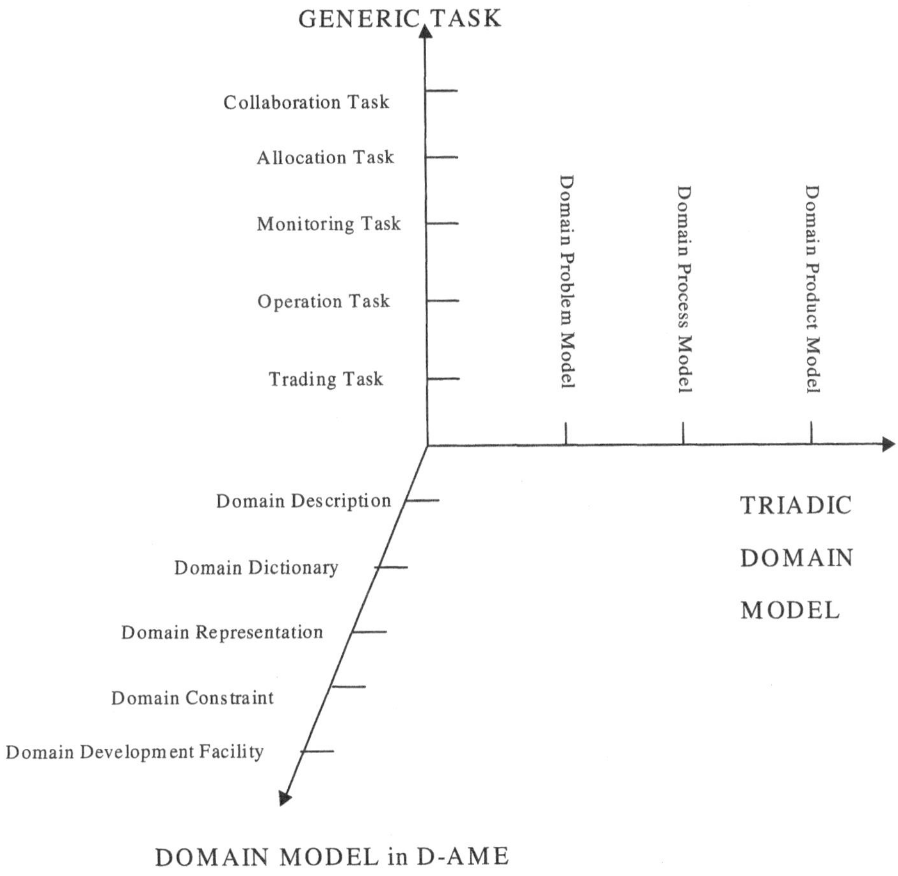

Figure 1.2 Triplex schema of concept for the domain analysis and modeling

The development teams or organizations have been adjusting and optimizing general development methods in order to fit into the target domains. Moreover, they have been designing and utilizing original or "domain oriented" methods for systems development. These methods can have various types of representations such as standard of the templates of requirements specifications or design specification, standards of various documentations, specification or program libraries, the group of software tools, and the principles for organization on development teams. These methods are unlikely necessarily shown outside of the systems development teams or organizations. It might be risky to apply general existing models or methods in the form of the state as it is without considering peculiar properties of the domains. The study on the adaptability of the assumption of general methods for the requirements analysis, design or implementation of target systems is important.

To use the word of "domain oriented" prompts to improve the degree of the recognition of reusing best or better practices of systems development. As a specified purpose, a domain analysis and modeling is not the passive or waiting process but the active process for accumulation, refinement, analysis, and modelling for the best or better practices.

As keys toward practical applications of "domain analysis and modeling" and "domain oriented systems development", the authors recognize the following three concepts: (1) D-AME framework, (2) Triadic Domain Model and (3) integration of models of Generic Tasks. This paper shows these key concepts. These concepts can be considered orthogonal and enable to organize diversified practices of domain oriented systems.

1.2 D-AME FRAMEWORK

A research working group of IPSJ SigSE (Information Processing Society of Japan, Special Interest Group of Software Engineering) had been studying the concept and applicability of domain analysis and modeling since 1993. This group summarized the schema of the domain analysis and modeling to provide a framework of domain oriented systems development and named it D-AME (the abbreviation of Domain Analysis, Modeling and Engineering).

Our D-AME research working group enumerate the purpose of domain analysis and modeling as follows:

- To reuse the past products such as software, software components and software specifications in various development stages.
- To systematize, clarify, improve, and create development process, methods, and tools appropriate for the domain.
- To collect, systematize, and educate development cases, know-how, experiences and methods in the development organization.

Main features of D-AME, which was described in a book written by D-AME research working group (Itoh, 1998), are summarized as follows (See Figure 1.3.):

- three processes in D-AME
- five components of domain model in D-AME
- interrelationships between three processes and five components of D-AME
- exhaustive coverage of domain model
- process model as domain model
- interrelationships between system lifecycle and five components of domain model

Our working group defines D-AME process as three processes consisting of (1) analysis process, (2) modeling process, and (3) engineering process. Throughout the D-AME process the domain model residing in the domain is specified. The domain model can be used during requirements analysis, design and implementation of systems, which are instances of the domain models. It should be noticed that none of these three sub-processes are limited to, respectively, requirements analysis, system design, and implementation in systems development.

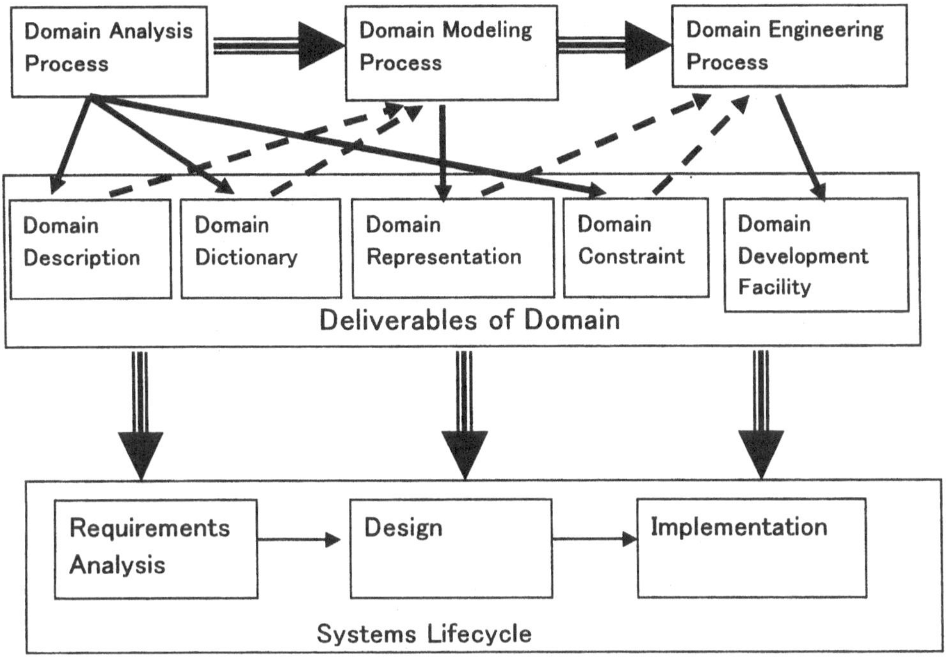

Figure 1.3 D-AME Framework

For example, a domain constraint, which is one of components of the domain model, is used not only in the implementation stage, but also in the system analysis or design stage. The three processes in D-AME are similar to those of Arango's Domain Engineering as in (Arango, 1989, 1991, 1995), which consists of (1) domain analysis, (2) infrastructure specification or domain modeling, and (3) infrastructure implementation or infrastructure construction.

Engineering process in D-AME means the construction of engineering environments for the systems development and reuse based on the recurrent practices in the domain oriented systems development. The engineering process in D-AME process produces engineering functions, tools, methods or environments appropriate for either requirements analysis, system design, or implementation in system instances belonging to the domain.

As the second feature, the domain model in D-AME consists of five components, i.e., (a) domain description, (b) domain dictionary, (c) domain constraint, (d) domain representation and (e) domain development facility. As the third feature, three processes of D-AME, i.e., analysis process, modeling process and engineering process, use these components as input and output. As the fourth feature, D-AME exhaustively enumerates actual examples of those five components of the domain model.

As the fifth feature, these actual examples include development process model, development methodologies, development tools, development environments as well as domain specific problem models and products models.

Using the five components of the domain model, appropriate characteristics and knowledge in the objective application domain can be used at various stages such as the system planning, the system requirements analysis, the system design, the system implementation, the system testing, and the system operations / maintenance during development.

The input for the first step "analysis process" in D-AME process includes various information and knowledge concerning the specific domain. This input should also include the objective and the purpose in D-AME process as follows:

1. Domain selection and identification of information sources
2. What can be reused in systems development?
3. Who can reuse the domain model?
4. At which phase analysis, design, or implementation can the domain model be reused?
5. Is the domain model for problem, process, or product?
6. What should be the domain development facility?

The output for the first step "analysis process" in D-AME process includes domain description, domain dictionary and domain constraint. The domain description specifies basic characteristics in the domain and itemizes, informal, or semi-formal fashion, all the information involved in the domain. The domain dictionary includes the terminology in the domain. The domain constraint includes the conditions, assumptions and limitations in developing system instances belonging to the domain.

The input for the second step "modeling process" in D-AME process includes the domain description and the domain dictionary. The output of the modeling process includes domain representation. The domain representation is a model in more formal or structured fashion. The modeling process produces the domain representation by characterization, abstraction, classification and generalization processes via the comparison of the similarities and differences between systems belonging to the domain.

The input for the third step "engineering process" in D-AME process includes the domain representation and the domain constraint. The output of the engineering process includes domain development facility. The domain development facility includes libraries, engineering tools engineering methods and engineering environments for the systems development and reuse based on the recurrent engineering practices in the domain oriented systems development. The domain development facility can be functions, tools, methods or environments appropriate for either requirements analysis, system design, or implementation in system instances belonging to the domain.

A narrower definition of a domain model can be used, corresponding to the domain representation above. The definition of domain model by D-AME process is broad. Of course, a domain model does not have to include all the above items, and the term "model" should not be defined in an overly mathematical way. Any, to some extent, formal representation such as, tables, languages, and charts, can be called a domain model in D-AME, or a domain representation as a domain model in a narrower sense.

The detail and examples of D-AME, e.g., are shown in (Itoh, 1997, 1998) and (Kumagai, 1998-a). The authors' examples are domains for facility reservation, facility management, facility monitoring, slip processing, DB application, hospital management, sales / rental / lease management, library management, OLTP real time system, plant control, collaboration system, etc. Examples by members of D-AME working group are domains for accounting system, manufacturing system, ATM for banking, telecommunication system, air-conditioning system, CAD for architecture, expert system, logistics, etc.

1.3 TRIADIC DOMAIN MODEL

The authors propose "Triadic Domain Model" (TDM) in (Itoh, 1994-a, -b, 1998). TDM consists of three constructs, i.e., Domain Problem Model, Domain Product Model and Domain Process Model. "Domain problem model" is a model of a set of problems to be systematized that include components with specific attributes, behaviors and interrelationships. "Domain Product Model" is a model of a set of solutions to be systematized such as specifications, program codes, algorithms and database schema that are solutions for the problems, In addition, "domain process model" is a model of the process by which the latter model is derived from the former model.

Domain Problem Model identifies problem components on their behaviors, properties and relationships by analysis of a class of problems. Domain Problem Model is described by the representation styles such as diagrams, forms, templates, scenario, fill-in-blank documents. Selection of representation styles depends on application domain. Terminology, data type, value range, component interrelationships, component dynamics, etc. depend on application domain. Domain Problem Model does not include methods / mechanisms on how to solve the problems.

Domain Product Model identifies specification / program components on their behaviors, properties and relationships by analysis of a class of system products. Domain Product Model includes solution methods / mechanisms by algorithms, data structures, computational resources, libraries, etc. During analysis and design steps, Domain Product Model is a model of specification. During programming step, Domain Product Model is a model of program.

Domain Process Model identifies typical activities and feasibility constraints by analysis of practices on a class of systems developments. When Domain Problem Model is described by a form representation, Domain Process Model specifies the standard ordering for filling the items of the form. This ordering is in partial ordering which is determined by domain heuristics. Domain heuristics also include recommendations, constraints, limitations and selection criteria on algorithms, data structures, computational resources, libraries, etc. to be adopted. Domain Process Model is the development process model that derives the Domain Product Model from the Domain Problem Model.

In order to perform systems development effectively on the basis of application domain, we present the Triadic Domain Model (TDM) based systems development environment. The main purpose is to integrate system specification, system design and generation for System.

1.4 INTEGRATION OF MODEL BY GERERIC TASK

The authors have analyzed important various domains such as the rental management domain, the sales management domain, the library management domain, the OLTP real time system domain, the plant control domain (Itoh, 1994-a, -b), the medical inspection domain (Itoh, 1997), and the railway scheduling domain (Kawabata, 1997), etc. , and have found each domain process model in reference to TDM.

The authors have recognized common tasks acknowledged across several domains. Such common tasks are found to perform in several different domains. For example, in the sales management domain, the inventory management performs the task of accepting order and delivery using slips and ledgers. The common task is also found in the plant control domain. In the sales management domain, the allocation task deals with the scheduling of personnel. In the plant control domain, it deals with the allocation of personnel for operation or inspection activities. We recognize these common tasks as generic tasks. When we analyze a domain, it is effective to recognize generic task at first. In order to validate the reusability of generic task beyond a particular domain, this chapter deals with prototyping cycle of generic tasks as a domain model to be reused.

The domain analysis by generic tasks can be used across several domains. For example, from the viewpoint of slip processing task as a generic task, we can analyze various domains by identifying slips, ledgers, personnel, and the interrelationships. Also, from the viewpoint of allocation task as a generic task, we can analyze various domains by identifying various kinds of stocks and delivering / shipping schedules.

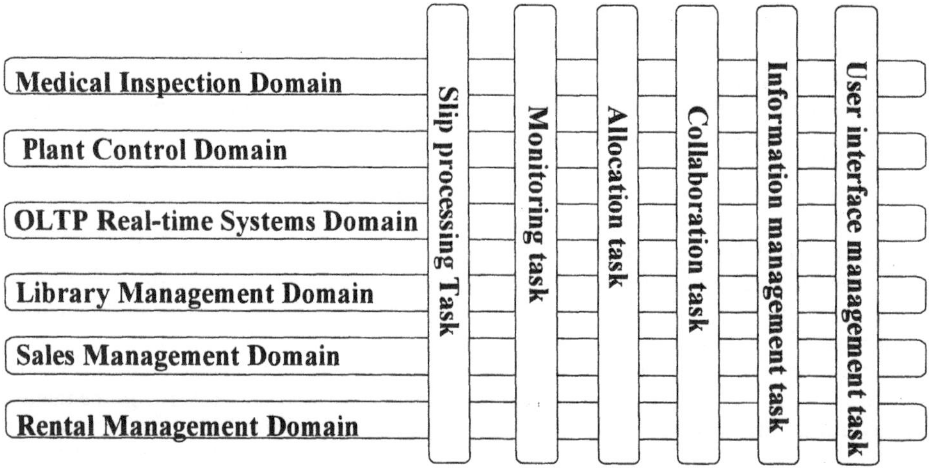

Figure 1.4 Generic Tasks and Domains

The authors recognized that the same kinds of tasks such as the slip processing task, the collaboration task, the allocation task, the monitoring task, the information management task, and the user interface management task, etc. are commonly performed across these domains. As an example, we obtained a domain model by the domain analysis for the cooperative work domain. The applicability of the domain model of common general works such as cooperative works range over different application areas widely. The authors call these tasks as generic tasks.

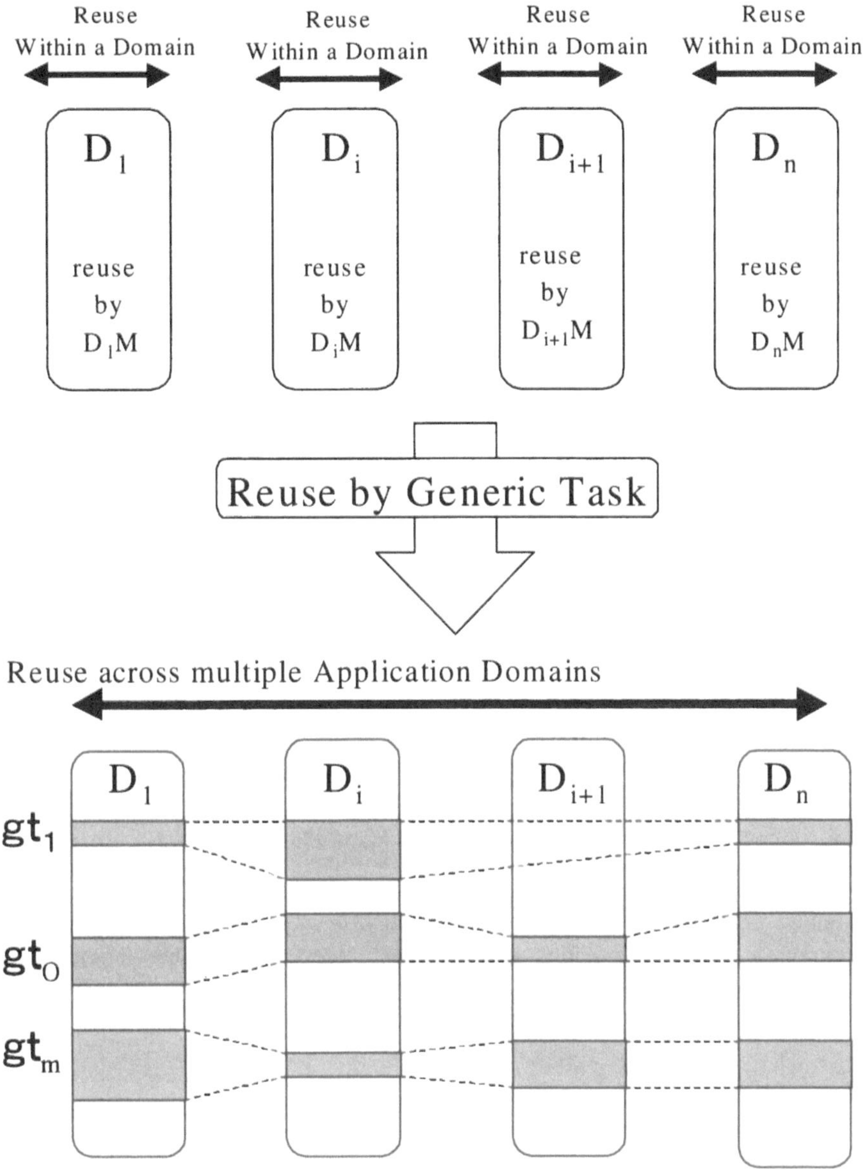

Figure 1.5 Reusing Domain Model by Generic Task

In general, it is difficult to reuse a domain model in another domain. Since domain model is often developed in some organizational context, sometimes it is not suitable for use in different organizational context. Issues for recycling domain model to different domains are:

- Criteria for judging neighbourhood domain in which domain model is reused
- Difference of terminology and /or business convention among domains

- Criteria for significance and superficiality of a domain in order to specify reusable part of domain

Recognition of the same kinds of tasks as generic tasks is important. The model of each generic task can be applied across the domains. Each domain can be considered as a combination of one or more generic tasks as shown in Figure 1.5. If the generic tasks are identified, the process models of the generic tasks accumulated in the past can be reused.

The authors obtain the terms used by the business / activity domain, and accumulate them. We clarify what kind of the generic tasks are done in the domain. The model of the generic tasks can be reused across two or more business / activity domains as shown in Figure 1.4. A domain model is represented as an integration of models of the generic tasks. The relationships of domains and generic tasks are similar to those in AI fields.

An example of generic task, collaboration task is shown in (Kawabata, 1998) (Kumagai, 1998-a, -b). Generic tasks can be modeled in terms of TDM.

1.5 APPROACH FOR APPLICATION

1.5.1 Various Tools

The authors have been applying the triplex scheme of key concepts to various domains. In the process of these applications, the authors have developed many tools that exemplify domain development facilities as shown in Table 1.1. Each of these tools has one or more generic tasks to be analyzed and designed, one or more charts to be used for analyzing and designing systems in domains, and one or more process models to be used. The detail of these tools are shown in (Kawabata, 1998, 2001, 2002-a, -b), (Kamimura, 2001) and (Kumagai, 1998-b).

The authors recognized, as process models, prototyping, terminology (ontology) extraction, collaboration analysis, state transition analysis, performance analysis, allocation analysis, information analysis, user interface elicitation, monitoring task analysis, and evolutional analysis. The evolutional analysis means the combination of analyses with two ore more process models.

1.5.2 NAVI

NAVI supports to analyze slip processing task as generic task. It uses tables as the charts to be used for analysis and design. It supports to advance prototyping process and term (ontology) extraction process as domain process model.

Academic affairs information systems domain is analyzed as an example of domain model. Prototyping of this domain is performed from the viewpoints of slip processing task. Prototyping process is regarded as a domain process model. Prototyping process by generic tasks can be reused across several domains.

Prototyping is an effective method for systems development. Prototyping uses the prototype of a product in order to verify specifications on theirs function, performance and cost at early stage in systems development. As the prototype can reduce misunderstandings or mistakes in early stage of development process, modification and re-planning of specifications can be expected to be reduced in latter stage.

Table 1.1 Various Development Facilities based on Triplex Scheme of Key Concepts for Domain Oriented Systems Development

		generic task	chart	process model
1	NAVI	slip processing	table	prototyping, terminology (ontology) extraction
2	MCM Editor	collaboration	MCM	collaboration analysis
3	CLM Editor	collaboration	CLM	collaboration / state transition analyses
4	MCM-IDEF0 Navigator	collaboration	MCM, IDEF0	collaboration / evolutional analyses
5	MCM-CLM Navigator	collaboration	MCM, CLM	collaboration / state transition / evolutional analysis
6	MCM-GPSS Navigator	collaboration	MCM, GPSS	collaboration / performance / allocation / evolutional analyses
7	MCM&CLM-ER Navigator	collaboration / information management	MCM, CLM, E-R	collaboration / state transition / information / evolutional analyses
8	IDEF0-ER Navigator	collaboration / information management	IDEF0, E-R	collaboration / information / evolutional analyses
9	Petri Net-ER Navigator	collaboration / information management	Petri Net, E-R	collaboration / information / evolutional analyses
10	usecase-class diagram-sequence diagram-usecase Navigator	use interface management	use case, class diagram, sequuence diagram	user interface elicitation / evolutional analysis
11	ER-SFC Navigator	collaboration / information management	E-R, SFC	collaboration / information / evolutional analyses
12	STD Triad Editor	monitoring	STD Triad	monitoring task / state transition analyses
13	ICCEE	collaboration / information management	MCM, CLM, E-R, IDEF0, GPSS	collaboration / prototyping / information / state transition / performance / allocation / evolutional analyses

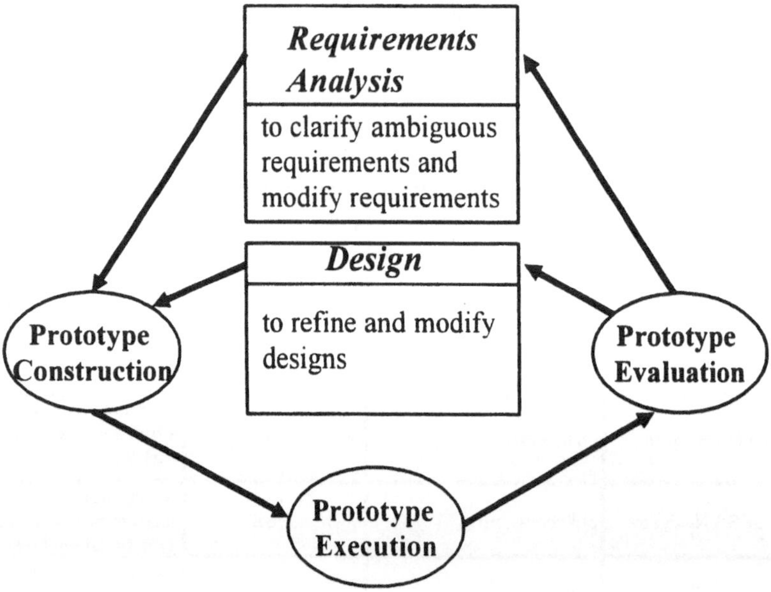

Figure 1.6 Prototyping Cycle

Prototype is constructed according to the prototyping cycle as shown in Figure 1.6. The prototyping cycle includes prototype construction, execution and evaluation.

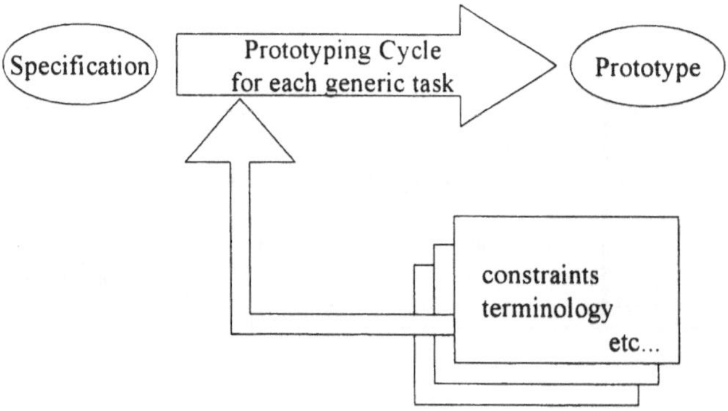

Figure 1.7 Reuse from the viewpoint of Generic Task

We regard a prototype as a domain product model and a prototyping cycle as a domain process model in TDM (Triadic Domain Model). Systems developer constructs prototype based on the specifications by customer. The developer validates and evaluates function, performance and cost by using prototype, and advances to more detailed specifications.

In Triadic Domain Model, we regard prototype as product model, the prototyping cycle as process model. In many cases, prototyping is performed based on the specific process for each domain. In this case, the domain process model and domain product model can be reused only in the target domain.

Using models such as the prototype and the prototyping cycle for each generic task enables to perform effective prototyping. By prototyping based on generic task, the prototyping cycle can be reused in different domain. Primary components and behaviour of generic task are considered to be same in each other in different domains. We can reuse the prototyping cycle for each generic task by linking domain library such as terminology, structure, knowledge and constraints for each domain as shown in Figure 1.7. The integration of library and the prototyping cycle help us to reduce efforts of prototyping. When we construct prototype for the other domain, the library for the domain will have to be obtained.

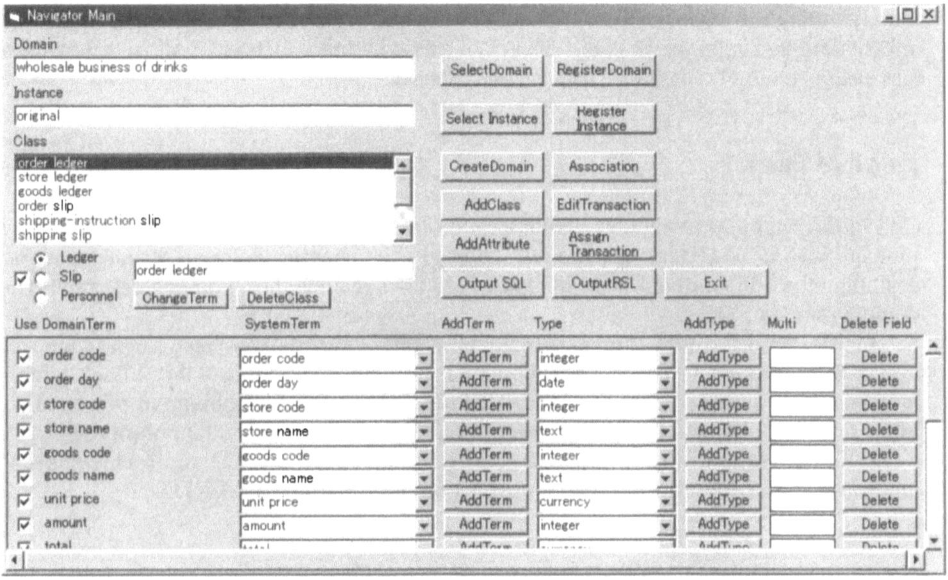

Figure 1.8 Definition of table

The authors implement the analysis process for the slip processing task as NAVI. By using NAVI, the authors can reuse analysis process for the slip processing task and accumulate new domain model in the library. NAVI supports to define transactions and data structures of personnel, ledger and slip. These can be accumulated in library as domain model or system instance. Users can perform prototyping easily by using this library. NAVI can outputs system instance as SQL form. Users can evaluate specifications by executing this SQL form on database software. By using NAVI, users can perform prototyping in a short period of time. Figure 1.8 shows the definition of the order ledger. The detail of slip processing task as a generic task is shown in (Kawabata, 1999).

1.5.3 MCM Editor

MCM Editor supports to analyze collaboration task as generic task. It uses Multi-Context Map (MCM) as the charts to be used for analyzing and designing the collaboration task. It supports to advance collaboration analysis process as domain process model.

Collaboration task is a process which involves diversified individual and / or organizations. Naturally, each participant in the process has different roles or perspectives in the process, while sharing a shared goal. Diversity of those roles or perspectives provides various contexts in the collaboration task. In other words, a collaboration task can be regarded as a collection of contexts. Each contexts switches one after another among the participants according to the differences of their perspectives. Specifying workflow which directs efforts of each participant is effective way to leverage their complementary skills and resources. In order to facilitate building the workflow, MCM and its construction process are presented. MCM can specify both information and material flows of the workflow. The construction process enables an incremental building of workflow (Kumagai *at el.*, 1998) for analyzing the collaboration task as shown in Figure 1.9. By using MCM, we can distinctively describe the flows of fact (information) and material recognized by the personnel. Who receives what information and material is clear. MCM can describe the difference of the information that each personnel recognize for the same material.

1.5.4 CLM Editor

CLM Editor supports to analyze collaboration task as generic task. It uses Collaborative Linkage Map (CLM) (Haswgawa *et al.*, 2000) as the charts to be used for analyzing and designing the collaboration task. It supports to advance collaboration analysis process as domain process model.

CLM specifies state transitions of personnel, materials or resources, which are important components in collaboration tasks as shown in Figure 1.10. Each state is represented in a circle. Those states of CLM are defined as a combination of Personnel-State Transition Diagram (P-STD), Material-STD (M-STD), and Collaboration-STD. (C-STD). P-STD shows state transitions of personnel and M-STD represents those of materials. C-STD describes the relationship's between P-STD and M-STD.

1.5.5 Bidirectional MCM-IDEF0 Navigator

Bidirectional MCM-IDEF0 Navigator supports to analyze evolutional task with collaboration task as generic task as shown in Figure 1.11. It uses MCM and IDEF0 constructs as the charts to be used for analyzing and designing the collaboration task. It supports to advance evolutional analysis process including collaboration as domain process model.

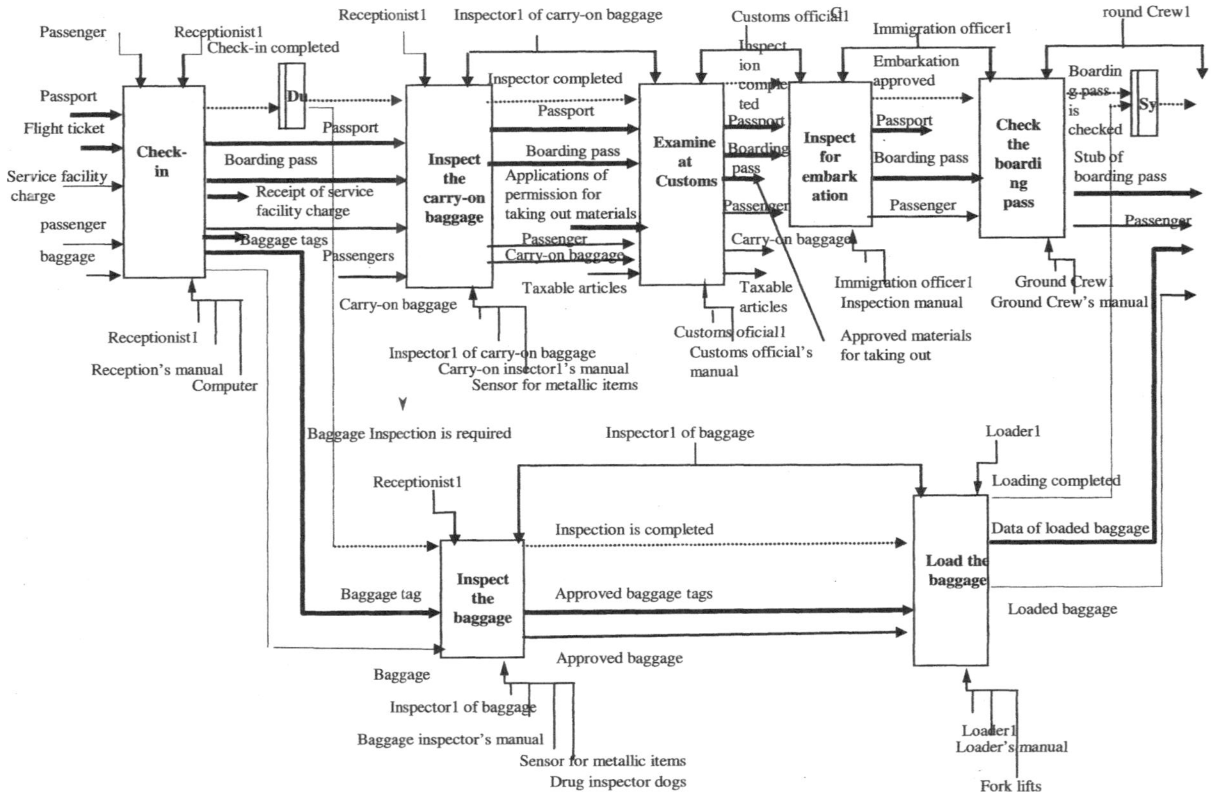

Figure 1.9 MCM for Airport

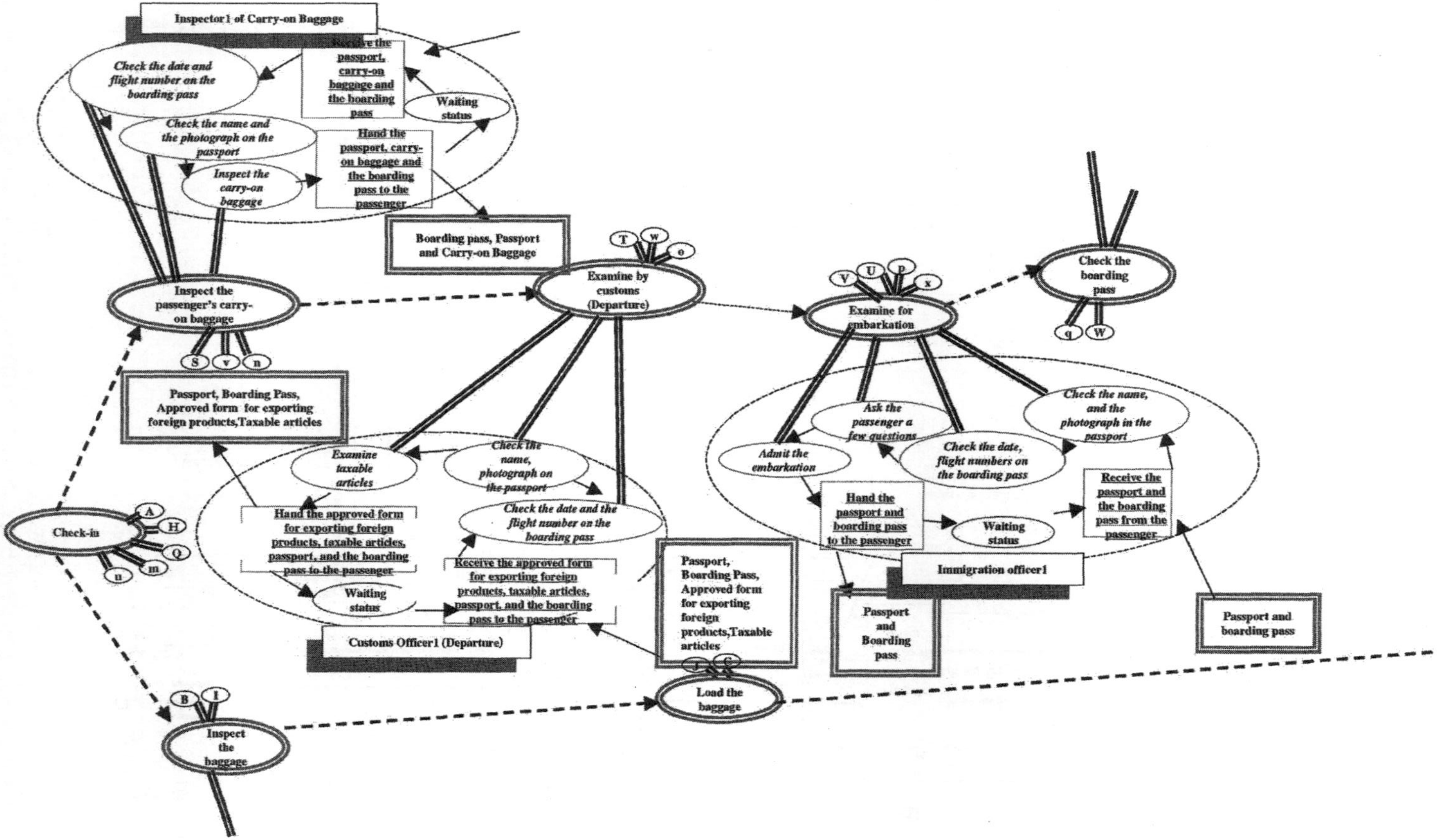

Figure 1.10 CLM for Airport

In IDEF0, a process is described by a set of activities. Everything such as data, information, etc. which relates to an activity is classified into ICOM (Input, Control, Output and Mechanism). Since the process of describing IDEF0 is not necessarily strict, the resulting flexibility is convenient to describe the systems at the analysis phase to grasp the overview, and it is easy to use for domain experts because the description is simple and easy to describe and understand. However, less-disciplined IDEF0 which includes ambiguity is likely to be obtained because of the flexibility. In particular, at the design phase, its description becomes too ambiguous for systems designers to proceed to detail design. In the case of analyzing the complex business process which includes collaboration, the classification of ICOMs is ambiguous because the meanings of things which relate to activities are different depending on the designer or engineer. In the collaboration analysis, analyzing the viewpoint and the position of the collaborator is important. Especially, it is necessary to describe the process distinguishing information flow and material flow in the task. Those meanings should be classified accurately. The method to classify the ICOMs accurately by which everyone can obtain the same result is necessary. The authors have developed the method to obtain the well-disciplined IDEF0, in which the viewpoint is strictly controlled and information and material are clearly distinguished.

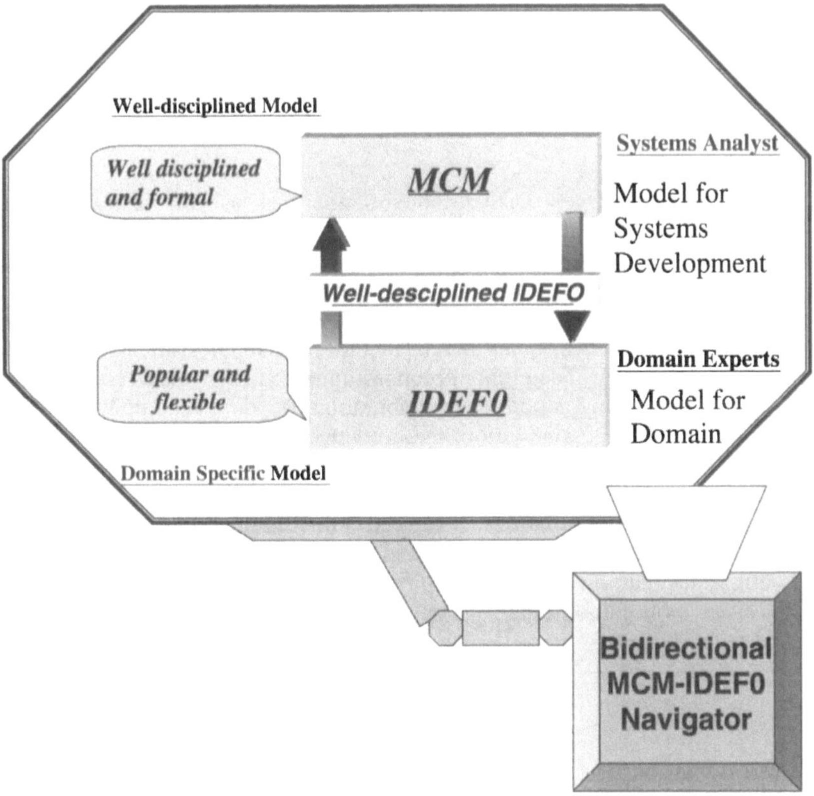

Figure 1.11 Process Supported by Bidirectional MCM-IDEF0 Navigator

By using MCM, the authors can describe the collaboration process distinguishing three kinds of flow such as those of information, material and token and using junctions, which tend to become ambiguous in IDEF0 descriptions. MCM is suitable for systems development because it has no ambiguity residing in IDEF0 description. However, MCM sometimes becomes difficult to understand particularly for business experts. The navigator supports bi-directional conversion between well-disciplined-IDEF0 and MCM. The navigator finds and shows ambiguous components and prompts user to add the MCM specific components such as perspectives to generate MCM. Thus, this conversion is semi-automatic process. The reverse conversion process from MCM to IDEF0 is automatic. IDEF0 can be obtained from MCM by eliminating information.

1.5.6 MCM-CLM Navigator

MCM-CLM Navigator supports to analyze collaboration task as generic task. It uses MCM and CLM constructs as the charts to be used for analyzing and designing the collaboration task. It supports to advance evolutional analysis process including collaboration / state transition analysis processes as domain process model.

This navigator produces the kernel of CLM from MCM. It prompts to refine the CLM in more detail. The kernel consists of the enumeration of Personnel-State Transition Diagram (P-STD), Material-STD (M-STD), and Collaboration-STD. (C-STD).

1.5.7 MCM-GPSS Navigator

MCM-GPSS Navigator supports to analyze collaboration task as generic task. It uses MCM and GPSS constructs as the charts to be used for analyzing and designing the collaboration task. It supports to advance evolutional analysis process including collaboration / performance / allocation analysis processes as domain process model.

For effective collaboration task, it is necessary to evaluate or estimate the performance of collaboration task quantitatively in the course of systems analysis in order to improve the efficiency. The authors propose a method to evaluate collaboration task quantitatively by converting a model of collaboration task described in MCM into a simulation model. By executing simulation, the bottleneck and the critical path among the activities which affects the performance can be detected, and resources can be reallocated to the activities.

A model described in MCM can be converted into a simulation model by adding information such as the number of personnel and equipment, the quantities of time length, information and material. We develop a navigator in Prolog, which converts a model of MCM into a simulation model, which can be executed in GPSS as shown in Figure 1.12.

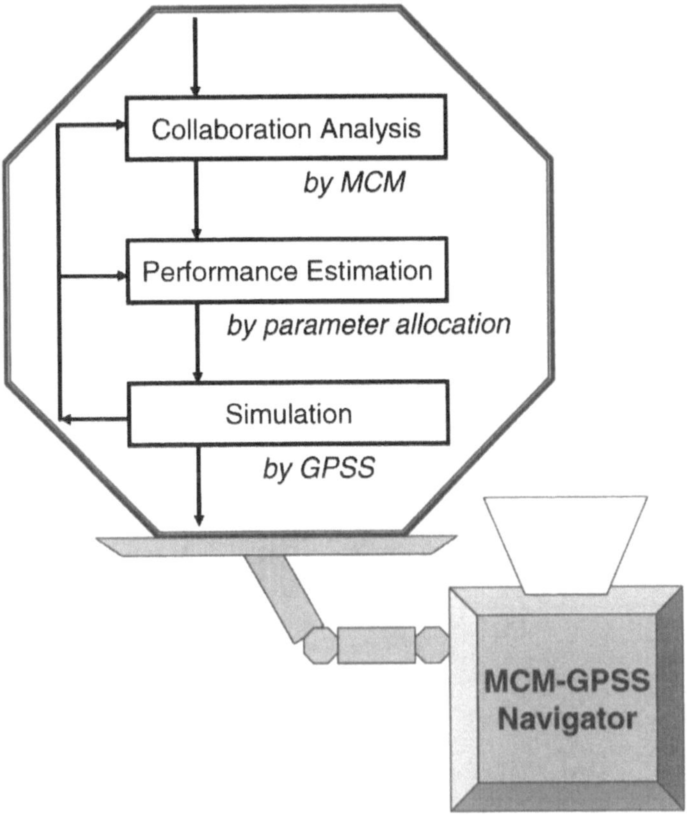

Figure 1.12 Process Supported by MCM-GPSS Navigator

1.5.8 MCM&CLM-ER Navigator

MCM&CLM-ER Navigator supports to analyze collaboration task and information managements as generic tasks. It uses MCM, CLM and E-R Diagram (Entity-Relationships Diagram) as the charts to be used for analyzing and designing the collaboration task and the information management task. It supports to advance evolutional analysis process including collaboration / state transition / information analysis processes as domain process model as shown in Figure 1.13. This navigator produces the kernel of E-R Diagram from MCM and CLM. It prompts to refine the E-R Diagram in more detail.

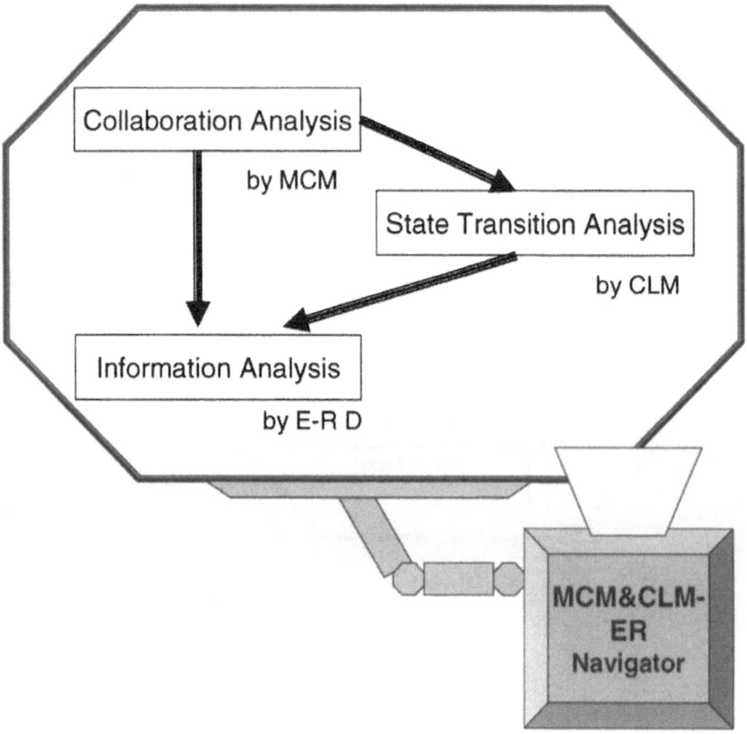

Figure 1.13 Process Supported by MCM&CLM-ER Navigator

1.5.9 IDEF0-ER Navigator

IDEF0-ER Navigator supports to analyze collaboration task and information managements as generic tasks. It uses IDEF0 and E-R Diagram (Entity-Relationships Diagram) as the charts to be used for analyzing and designing the collaboration task and the information management task. It supports to advance evolutional analysis process including collaboration / information analysis processes as domain process model. This navigator produces the kernel of E-R Diagram from IDEF0. It prompts to refine the E-R Diagram in more detail.

1.5.10 Petri Net-ER Navigator

Petri Net-ER Navigator supports to analyze collaboration task and information managements as generic tasks. It uses Petri Net and E-R Diagram (Entity-Relationships Diagram) as the charts to be used for analyzing and designing the collaboration task and the information management task. It supports to advance evolutional analysis process including collaboration / information analysis processes as domain process model as shown in Figure 1.14. This navigator produces the kernel of E-R Diagram from Petri Net. It prompts to refine the E-R Diagram in more detail.

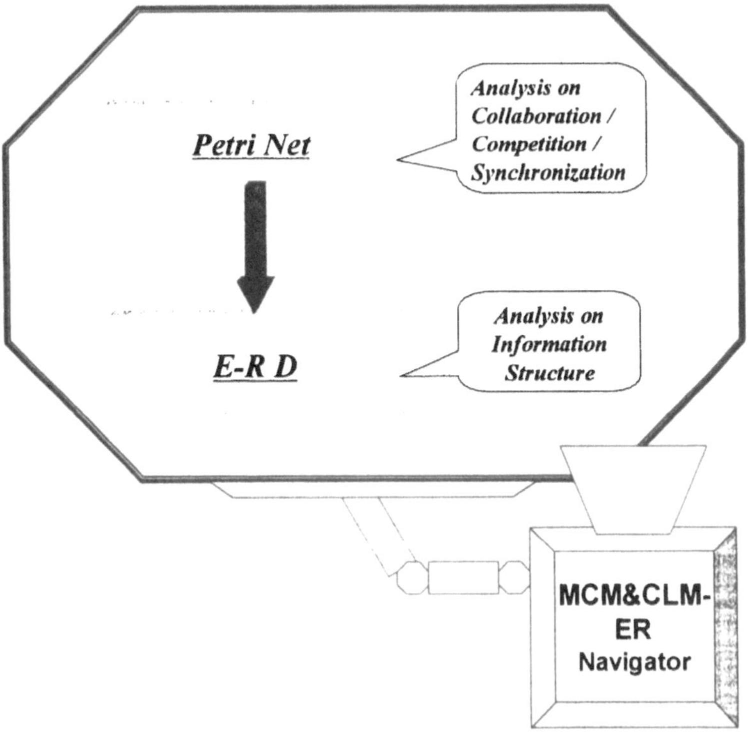

Figure 1.14 Process Supported by Petri Net-ER Navigator

Petri Net is a chart representation which represents synchronization and competition of resources such as material, information, personnel and server. Petri Net consists of transition, place, ark, and token. Place represents the states of starting and finishing the activity. Transition represents event which starts next activity. These two are connected by ark. Token is in place. Personnel who receive services such as guests is described as transaction. The token flow represents transaction flow. Various servers, which provide services such as personnel and machine, are described as place. When the token is in the place of server, the server is available. When the server which does the service is available, the arriving transaction can receive service. It is shown by the transition which represents synchronization. The servers and transactions are connected with the same transition. The content of service is written as a label beside transition. The entities should be enumerated as the main components of the system. In Petri Net there are places which represent the transaction and the server.

1.5.11 Navigator for User interface

This navigator supports to analyze user interface managements as a generic task. It uses use case, class diagram, and sequence diagram as the charts to be used for analyzing and designing the interface management task. It supports to advance evolutional analysis process including user interface elicitation as domain process model. This evolutional analysis means that the user interface for objective system can be obtained by serially constructing use case, class diagram, sequence diagram and use case. The final use case is a result of user interface elicitation process.

1.5.12 ER-SFC Navigator

ER-SFC Navigator supports the collaboration task as generic task. It uses E-R Diagram and SFC (Sequential Function Chart) as the chart to be used for analyzing and designing the collaboration task (Kumagai *et al.*, 1999). It supports to advance collaboration / information / evolutional analysis processes as domain process model.

Collaboration process is a complex process, which involves various kinds of persons or organizations. The authors analyzed the domain with reference to the TDM (Triadic Domain Model). TDM specifies a domain with domain problem model, domain product model, and domain process model. The domain problem model is identified by three kinds of elements – collaborator, its work items, and shared data. The authors set the domain product model as a combination of two representations – SFC (as shown in Figure 1.16) and E-R Diagram (as shown in Figure 1.15), which best describes the domain. The domain process model is a process for inducing those charts from the domain problem model. A navigator prototype is developed to implement the domain process model, which enables effective domain analysis and modeling practices and prompts reuse of the domain models to neighboring domains as shown in Figure 1.17.

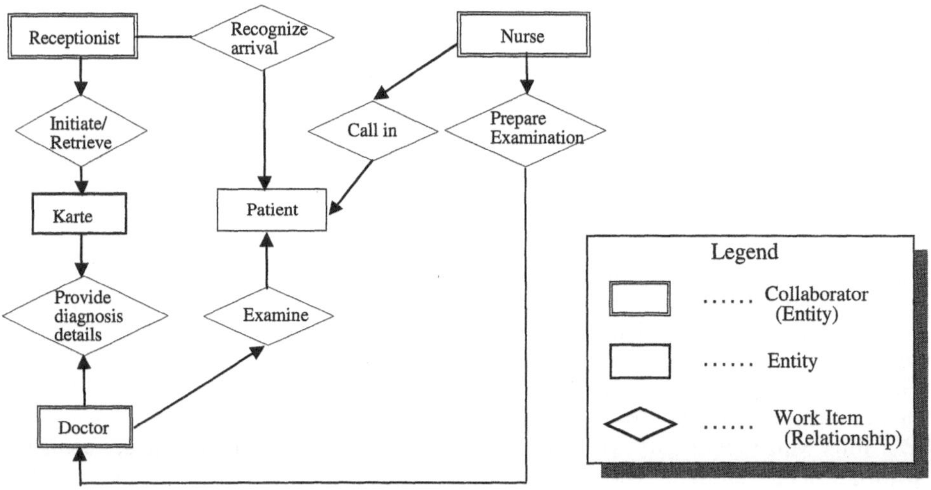

Figure 1.16 E-R Diagram for Medical Domain

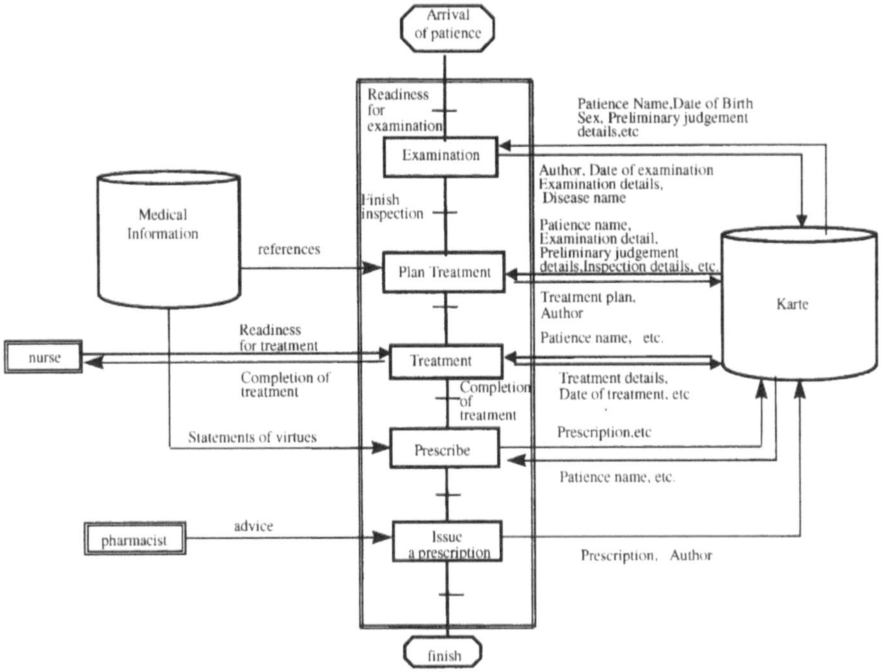

Figure 1.16 SFC for Medical Domain

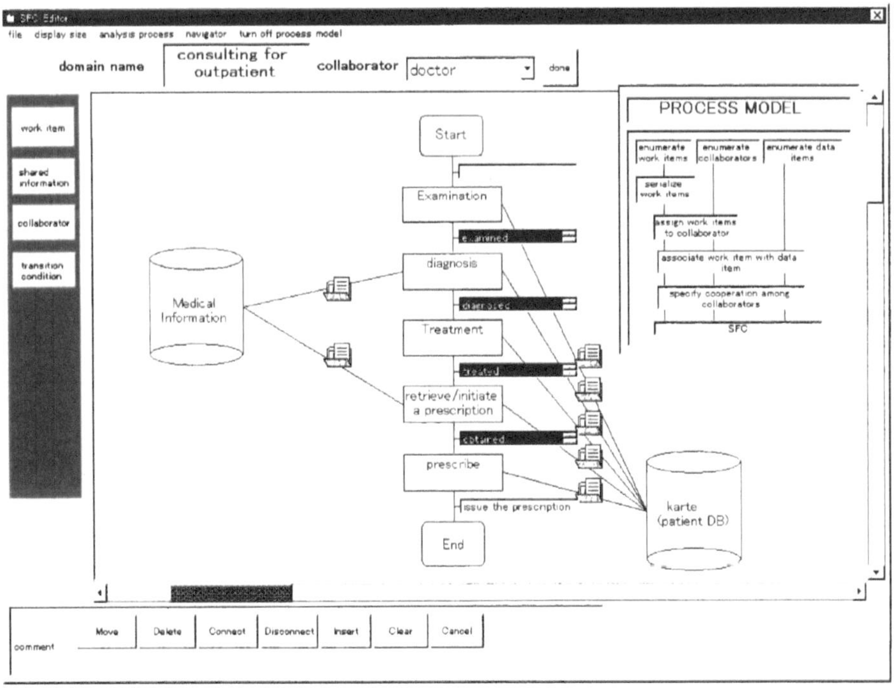

Figure 1.17 Navigator for Analysis

1.5.13 STD Triad Editor

STD Triad Editor supports the monitoring task as generic task. It uses STD Triad constructs as the chart to be used for analyzing and designing the monitoring task. It supports to advance monitoring / state transition analysis processes as domain process model.

STD Triad was devised by Kishima, S. and Itoh, K. in (Itoh, 1994-a). The application domain of STD Triad is a plant control domain. STD Triad is a State Transition Diagram (STD) that has three state groups, i.e., state groups by nature dynamics, by artificial enforcement, and by heuristic threshold, as shown in Figure 1.18. The first state group means a collection of states of the objective plant. The second one means a collection of states of the controller that senses and actuates the plant. The third one means a collection of states that second state one judges the conditions of first one.

1.5.14 ICCEE

ICCEE supports to analyze collaboration task as generic task. It uses MCM, CLM, E-R Diagram (Entity-Relationships Diagram) as the charts to be used for analyzing and designing the collaboration task (Kumagai *et al.*, 2000, Kawabata *et al.*, 2002-a). It supports to advance evolutional analysis process including collaboration / prototyping / information / state transition / performance / allocation analysis processes as domain process model.

Figure 1.19 shows the overall architecture of an integrated environment. In the integrated environment, collaboration interface model is the nucleus of the context base communications among the collaborators. The integrated environment is called ICCEE (Integrated Collaborative & Concurrent Engineering Environment).

Multi-Context map (MCM) is a description method according to the collaboration interface model. The collaboration task can be analyzed and designed in terms of function. MCM can be converted from IDEF0 which is widely used as standardized modeling technique.

Because the method of describing IDEF0 tends to include ambiguity in the description, and do not distinguish neither the material nor information clearly. It is not possible to convert the IDEF0 to MCM which distinguishes material flow and information flow. So, well-disciplined IDEF0 for collaboration task analysis/design is described by introducing the discipline based on the collaboration interface model. Because the well-disciplined IDEF0 distinguishes the material and information, it is possible to convert it to MCM semi-automatically.

By IDEF0 and MCM, the collaboration task can be analyzed in terms of function. To analyze the collaboration task in terms of performance, simulation is executed. MCM is converted into the simulation model for performance evaluation/estimation written with GPSS by giving MCM the amount of the processing time and the number of resource. The model enables to do performance evaluation/estimation of the collaboration task.

MCM is used to capture the whole of the collaboration task. Collaborative Linkage Map (CLM) is used to analyze state transitions of the resources such as collaborators which relate to the collaboration task, the material, and information. The description by MCM and CLM is complementary.

E-R Diagram for designing the data base can be extracted from CLM. The prototype and DB of the system which supports the collaboration task can be constructed with MCM, CLM, and E-R Diagram. The prototype is implemented with WWW server,

DB server, Mail server, and Token Flow Control server. It is possible to evaluate the system in terms of function by using the prototype which is executable on PC.

The system prepares windows for each collaborator. The necessary information is displayed in each window on client PC. When activity finishes, the collaborator inputs the result in the window. Windows for the collaborators and the token flow and information flow are designed from MCM. Information displayed in window, and inputted by the collaborator is designed by extracting E-R Diagram from CLM.

The authors proposed a method for analyzing a manufacturing system in term of collaboration (Hasegawa *et al.*, 2001). It is important for the efficient manufacturing system to capture materials and information shared among collaborators / equipments and distinguish between an internal setup operation and external setup operation. The authors applied MCM and CLM in order to capture collaboration among operations and information such as the change of the equipment and the setup operation to change over the machine activity in manufacturing system. The authors employ MCM for the description of a whole workflow in manufacturing system and CLM for the description of state transitions of collaborators, equipments and materials/information in the collaborative manufacturing system. MCM specifies the feasibility of concurrency in the collaboration as well as contextual switches among individuals or organizations. CLM describes state transitions of resources.

1.5.15 Detection and Correction of Specification Errors for Various Charts

The authors use many kinds of chart representations for systems analysis as shown above (Kawabata *et al.*, 2000). The chart representations should be described in accordance with the rules that are characteristic for each chart representation. The chart representations should be described correctly. The authors provide a method and its implementation to correct errors for chart representations in the course of systems analysis. The authors analyze chart representations with errors. The errors are classified into five types, (1) missing of essential components, (2) omission of components, (3) redundancy, (4) inconsistency, (5) ambiguity. Based on the classification, the authors develop a tool to support the analysis process for generic tasks. Generic task is a class of processes that can be observed across multiple application domains. To define requirements to a generic task oriented tool for the error correction, the authors analyze the five types of errors included in chart representations for collaboration task and trading task. The tool checks essential components and their inter-relationships in chart representations during analysis process for these two generic tasks. The tool suggests possible alternatives or candidates of valid components and inter-relationships by the evaluation on chart representations.

1.6 CONCLUDING REMARKS

An approach for domain oriented systems development is shown by a triplex schema of concept - D-AME framework, Triadic Domain Model, and Generic Task. Various application domains have been analyzed. Many types of domain development facilities have been developed.

This chapter is written by modifying and upgrading the early versions of articles in (Itoh, 1999).

For cooperation in development of tools shown in section 1.5, the authors thank Kengo Nakagawa, Yuji Senuma, Jun Maruyama, Miho Araki, Shinichi Tokuda, Naofumi

Kato, Naokazu Someya, Kiyofumi Inoue, Kan Tonami, Fumihiko Harada, Takahiro Kanbe, Daisuke Fukushima, Jiro Shinkai, Akira Shibata, Tatsuya Kakumu and Toshimitsu Hirai.

BIBLIOGRAPHY

Arango, G. 1989, Domain Analysis - From Art Form to Engineering Discipline, in *Proc. of 5th IWSSD*, , pp. 152-159.

Arango, G., Prieto-Diaz, R., 1991, Introduction and Overview: Domain Analysis Concepts and Research Directions, pp.9-26, in *Domain Analysis and Software Systems Modeling (R. Prieto-Diaz and G. Arango ed.)*, IEEE.

Arango, G., 1995, Domain Analysis, in *Marciniak, J.J. ed.: Encyclopedia of Software Engineering*, pp.424-434, John Wiley and Science.

Hasegawa, A., Kumagai, S., Itoh, K., 2000, Collaboration Task Analysis by Identifying Multi-Context and Collaborative Linkage, Journal of Concurrent Engineering Journal, Volume 8, Number 1/2 (March/June 2000).

Hasegawa, A., Kumagai, S, Itoh, K., 2001, Collaboration Engineering analysis for Manufacturing System, in *Proc. of CE2001*.

Itoh, K., Kishima, S., Tamura, Y., 1994-a, Systems Integration on Specification, Design and Generation of Reactive System - Triadic Domain Model-Based Approach –, in *Proc. of IEEE ICSI'94 (3rd International Conference on Systems Integration)*, pp.1082-1095.

Itoh, K., Kishima, S., Tamura, Y., 1994, Triadic Domain Model-Based Development of Software Systems, in *Proc. of IEEE ICSR'94 (3rd International Conference on Software Reuse)*.

Itoh, K., Kishima, S., Tamura, Y., Hirota, T., Hiroyuki Yoshida, ed., 1995, Domain analysis and modeling: Advanced Method for Software Development and Reuse, Kyouritsu Syuppan, Tokyo, Japan, in Japanese.

Itoh, K., Shinkai, J., Kumagai, S ., Kishima, S., 1997, A Method for Domain Analysis and Modeling of Cooperative Work, in *Proc. of SCS CEE'97 (Concurrent Engineering in Europe)*, pp.37-44.

Itoh, K., Hirota, T., Kumagai, S., Yoshida, H., 1998, Domain Oriented Systems Development: Principles and Approaches, Vol.1 of the series of Advanced Information Processing Technology, Gordon and Breach Science Publishers.

Kawabata, R., Itoh, K., Kumagai, S., 1999, Prototyping based on Domain Analysis and Modeling, in *Proc. of 4th IDPT (World Congress on Integrated Design and Process Technology)*.

Kamimura, M., Kato, N., Kawabata, R., Kumagai, S., Itoh, K., 2001, Extracting E-R models from Collaboration Analysis Methods, MCM and CLM, in *Proc. Humacs 2001 in ER2001*.

Kawabata, R., Tamura, Y., Itoh, K., Tomii, N., 1997, Concurrent Work Specification for Co-related Assignment Tasks with Domain-oriented Abstraction, in *Proc. of SCS CEE'97 (Concurrent Engineering in Europe)*, pp.133-140.

Kawabata, R., Itoh, K., 1998, System Analysis by Navigator for Collaboration Task, in *Proc. of SCS CEE'98 (Concurrent Engineering in Europe)*.

Kawabata, R., Itoh, K., Kumagai, S., 1999, Prototyping based on Domain Analysis and Modeling, in *Proc. of 4th IDPT (World Congress on Integrated Design and Process Technology)*.

Kawabata, R., Kumagai, S., Itoh, K., 2000, A Navigation Method for Detection and Correction of Specification Errors on Chart Representations, in *Proc. of 5th IDPT*.

Kawabata, R., Inoue, K., Kumagai, S., and Itoh, K., 2001, A Two-Faceted Systems Specification Method toward Collaborative Systems Engineering, in *Proc. of ICSSEA*.

Kawabata, R., Hasegawa, A., Kumagai, S., Itoh, K. 2002-a, Integrated Collaborative & Concurrent Engineering Environment : ICCEE, in *Proc. Of 6th IDPT*, (to appear).

Kawabata, R., Tokuda, S., Kamimura, M., Hasegawa, A., Kumagai, S., Itoh, K., 2002-b, Effective Systems Analysis by Reusing Different Kinds of Chart Representations, in *Proc. of 6th IDPT*, (to appear).

Kumagai, S., Hasegawa, A., Kawabata, R., and Itoh, K., 1998-a, A Domain Model for Collaborative Work using Multi-Context Map, in *Proc of IDPT'98*.

Kumagai, S., and Itoh, K., 1998, Designing Collaborative Work in IDEF0 using Interface Model, in *CERA Journal*.

Kumagai, S., Hasegawa, A., Kawabata, R., Itoh, K., 1998, Describing Collaboration Task using Muti-Context Map, *in proc. of CE'98*.

Kumagai, S., Shinkai, J. and Itoh, K. 1999, Implementing Domain Process Model as a Navigator for Cooperative Work Domain, in *SDPS Journal*, Vol.3, No.3, pp.53-74.

Kumagai, S., Kawabata, R., Hasegawa, A., Itoh, K., 2000, Integrated Environment for Collaboration Engineering by Collaboration Interface Model, *in proc. of 5th IDPT*.

Prieto-Diaz, R., 1987, Domain Analysis for Reusability, in *Proc. of COMPSAC'87*, " pp.23-29.

Prieto-Diaz, R., Arango, G., 1991, Domain Analysis and Software systems Modeling, IEEE Computer Society Press, California.

Tracz, W., 1992, Domain Analysis Working Group Report, First Int. Workshop on Software Reusability", in *ACM SIGSOFT SE. Notes*, Vol.17, No.3, pp.27-34.

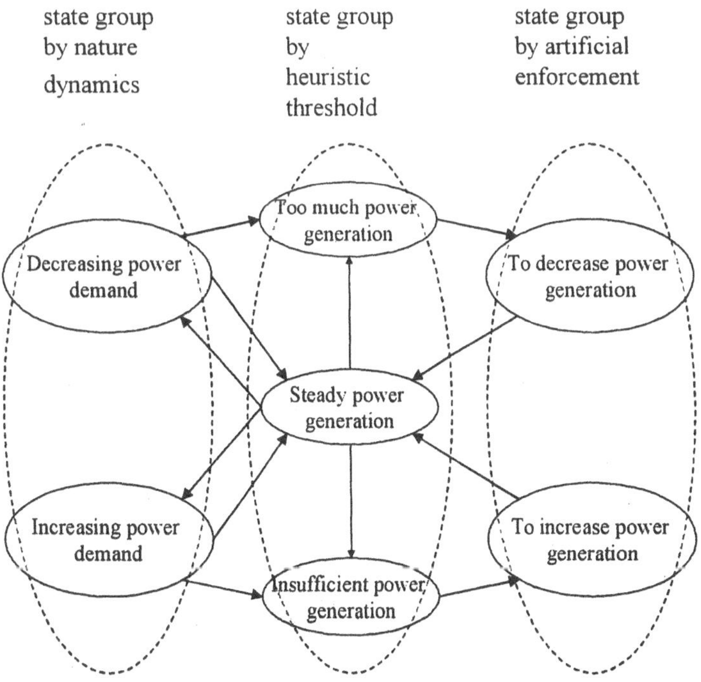

Figure 1.18 STD Triad for Power Generation

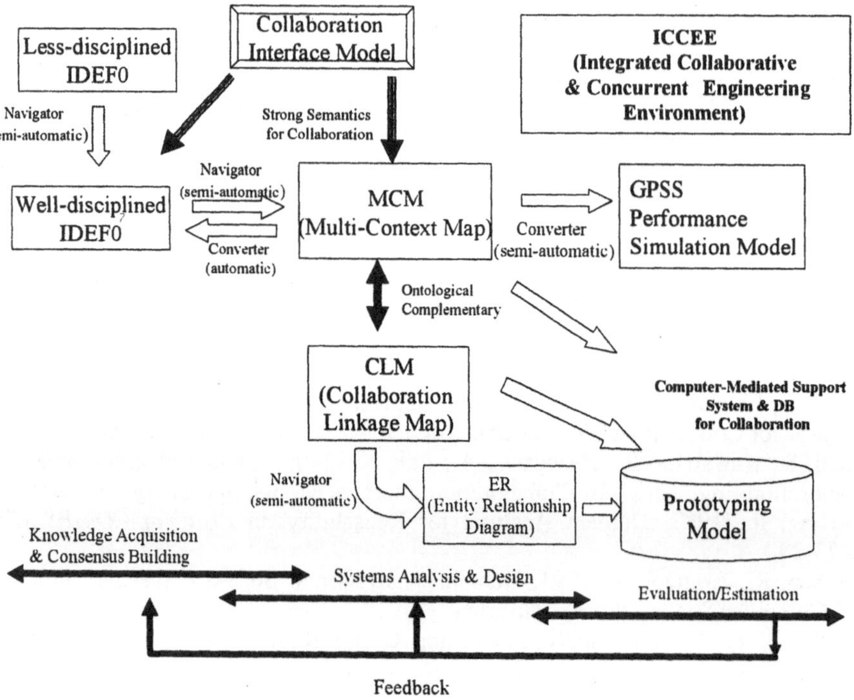

Figure 1.19 ICCEE

CHAPTER 2

Feature Oriented Product Line Software Engineering: Principles and Guidelines

Kyo C. Kang

Kwanwoo Lee

Jaejoon Lee

Pohang University of Science and Technology

Sajoong Kim

Korea IT Promotion Agency

ABSTRACT

Product line software engineering (PLSE) is an emerging software engineering paradigm, which guides organizations toward the development of products from core assets rather than the development of products one by one from scratch. To be able to engineer application software for a product line, the application domain of the product line must first be understood in terms of application family's commonality and variability. Since the feature-oriented approach to domain analysis (FODA) was introduced in 1990 by the Software Engineering Institute, many domain engineering and product line engineering methods for software reuse have adopted the technique to support the feature-oriented commonality and variability analysis. FORM (Feature-Oriented Reuse Method), for instance, extended FODA to include architecture design and object-oriented component development. This method has been applied to several industry application domains, including elevator control systems, electronic bulletin board systems, yard inventory systems, and PBX, to create product line software engineering environments and software assets.

This paper introduces the FORM method focusing on engineering principles and guidelines used for feature modeling, architecture design, and component development. Yard inventory systems are used to demonstrate the feature-oriented product line software engineering.

2.1 INTRODUCTION

Product line software engineering (PLSE) is an emerging software engineering paradigm, which guides organizations toward the development of products from core assets rather than the development of products one by one from scratch. Two major activities of

PLSE are core asset development (i.e., product line engineering) and product development (i.e., product engineering) using core assets. (Details of the product line software engineering framework can be found at http://www.sei.cmu.edu/plp/framework.html.)

The paradigm of developing core assets for application development has been called domain engineering (DE), in which an emphasis is given to the identification and development of reusable assets from an application "domain" perspective. Product line software engineering is similar to domain engineering in that they both attempt to exploit commonalities to build reusable core assets. However, PLSE differs from DE in that PLSE is founded on marketing. In PLSE, a product plan that specifies target products and their features from a market analysis is the primary input. Fielding products with features that the market demands in a timely manner and then evolving those products as the market evolves is the major driving force in the asset development in PLSE. Therefore, the scope of analysis and development in PLSE can be narrower and more focused than in DE. However, most engineering techniques used in DE can be applied in PLSE as both paradigms attempt to build flexibility and reusability into core assets.

In order to develop reusable core assets for a product line, PLSE must have an ability to exploit commonality and manage variability. A feature oriented approach to commonality and variability analysis has been used extensively for DE and PLSE both in industry (e.g., Italia Telecom (Vici *et al.*, 1998), Northern Telecom (Zalman, 1996), Hewlett Packard Company (Griss *et al.*, 1998; Griss, 2000) and Robert Bosch GmbH (Hein *et al.*, 2000)) and academia (e.g., Software Engineering Institute (Cohen *et al.*, 1991; Krut *et al.*, 1996), Technical University of Ilmenau (Czarnecki *et al.*, 2000), and Pohang University of Science and Technology (Kang *et al.*, 1998; Kang *et al.*, 1999; Lee *et al.*, 2000a; Lee *et al.*, 2000b), since the idea of feature oriented analysis (i.e., FODA (Kang *et al.*, 1990)) was first introduced in 1990 by the Software Engineering Institute. Many domain engineering methods for software reuse have adopted the technique to support the domain commonality and variability analysis. ODM (Simos *et al.*, 1996), FeatureRSEB (Griss *et al.*, 1998), FODAcom (Vici *et al.*, 1998), and DEMRAL (Czarnecki *et al.*, 2000) are examples of those.

We also extended FODA into FORM (Feature-Oriented Reuse Method) to support architecture design (Kang *et al.*, 1998) and object-oriented component development (Lee *et al.*, 2000a). This method has been applied to several industrial application domains, including electronic bulletin board systems (Kang *et al.*, 1998), PBX (Kang *et al.*, 1999), elevator control systems (Lee *et al.*, 2000a; Lee *et al.*, 2000b), and yard inventory systems, to create product line software engineering environments and software assets.

In this paper, we focus on the principles and guidelines used by FORM for product line software engineering with an example of yard inventory systems. As a starting point, section 2.2 describes FORM activities and their inputs/outputs for the product line asset development. The principles and guidelines used in each activity are then discussed in sections 2.3, 2.4, and 2.5. Section 2.6 demonstrates an application of the principles and guidelines to the product line engineering of yard inventory systems. Finally we conclude this paper with a discussion of lessons learned from our experience in section 2.7.

2.2 PRODUCT LINE ASSET DEVELOPMENT ACTIVITIES

The core concept of FORM lies in the analysis of product features and the use of these features to develop reusable and adaptable artifacts (e.g., architectures and components) for a product line. In this section, we provide an overview of the asset development aspect of the FORM method in terms of method activities and their inputs/outputs. (See

Figure 2.1 for FORM activities and their relationships.) FORM starts with feature modeling to discover, understand, and capture commonalities and variabilities of a product line. Feature modeling is to capture commonality and variability in terms of product features (i.e., capability, system environment, domain technology, and implementation technique features) and organize them into a model, which serves as a basis for deriving reusable and adaptable artifacts.

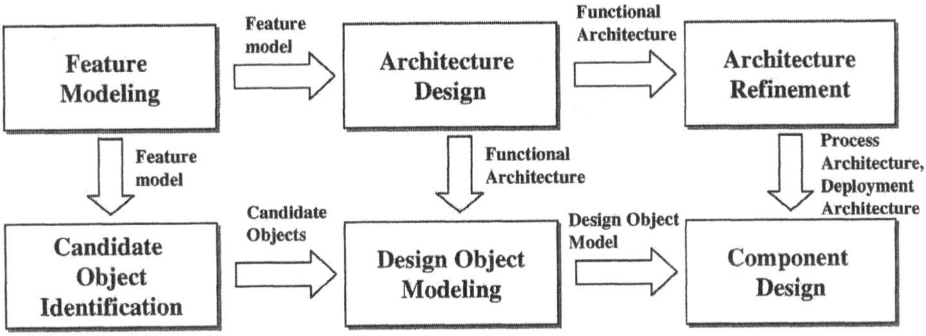

Figure 2.1 Asset development activities of FORM

The architecture design activity allocates features to architectural components and specifies data and control dependencies between architectural components. The result is a "functional architecture[1]." Architects then refine the functional architecture into "process and deployment architectures[2]" by considering whether to replicate or separate each component, allocating components to concurrent processes and network nodes, and defining methods of interactions between components. During the architecture design and refinement, architects must make design decisions considering critical quality attributes that are sensitive to architectural changes, so that these decisions are inherited to the component design preserving the quality attributes already explored during the architecture design.

For the component design activity, candidate objects, which are basic building blocks to create large-scale reusable components, must first be identified based on the feature model. A design object model is then developed based on the functional architecture, candidate objects, and other necessary information such as product line requirements, commercial off-the-shelf (COTS) components, and design patterns that are relevant to the product line.

The component design activity refines the process and deployment architectures into concrete components (i.e., modules) using the design object model with considerations of feature interactions, feature delivery methods, feature binding time, quality attributes, reusability, etc. The results of this activity are product line architectures and asset components.

Some of the modeling principles and guidelines that have emerged from the application of the FORM method to industrial systems are discussed in the following sections.

---

1 The functional architectural describes a system in terms of its functional elements and relationships between them.

2 The process architecture represents a concurrency structure in terms of concurrent processes (or tasks) to which functional elements are allocated; the deployment architecture shows an allocation of processes to hardware resources.

2.3 FEATURE MODELING GUIDELINES

Understanding commonality and variability is essential for developing reusable core assets for a product line. Feature modeling is an effective and efficient method for identifying and organizing commonalities and variabilities among the products of a product line. This section explores some of guidelines for feature modeling, especially in the context of product line analysis. Details of feature modeling can be found in (Kang *et al.*, 1990; Kang *et al.*, 1998; Kang *et al.*, 1999; Lee *et al.*, 2000a; Lee *et al.*, 2000b).

Features are any prominent and distinctive aspects or characteristics that are visible to various stakeholders (i.e., end-users, domain experts, developers, etc.). In our method, we provide feature categories as a feature identification framework:

- Capability features: These features are characterized as distinct services, operations, or non-functional aspects of products in a domain. Services are end-user visible functionality of products offered to their users in order to satisfy their requirements. They are often considered as marketable units or units of increment in a product line. Operations are internal functions of products that are needed to provide services. Non-functional features include end-user visible application characteristics that cannot be identified in terms of services or operations, such as presentation, capacity, quality attribute, usage, cost, etc.
- System environment features: These features include product line contexts, such as computing environments and interfaces with different types of devices and external systems. Protocols used to communicate with external systems are also classified as environment features.
- Domain technology features: These features are domain specific technologies that domain analysts or architects use to model specific problems in a domain or to "implement" service or operation features. Domain-specific theories and methods, analysis patterns, and standards and recommendations are examples. Note that these features are specific to a given domain and may not be useful in other domains.
- Implementation technique features: These features are more generic than domain technology features and may be used in different domains. They contain key design or implementation decisions that may be used to implement other features (i.e., capability and domain technology features). Communication methods, design patterns, architectural styles, ADTs, and synchronization methods are examples of implementation techniques.

In addition to the above feature identification guidelines, the following guidelines must always be kept in mind during the feature modeling activity.

Analyze terminologies used in the product line to identify features. We often see that domain experts in mature or stable domains use standard terms to communicate their ideas, needs, and problems. For example, 'call-forwarding' and 'call-back' are commonly used terminologies in the telephony domain. Identification of commonly used terms among domain experts can expedite feature modeling. In an immature or emerging domain, however, domain experts often use different terminologies for the same meaning or the same terminology with different meanings. Thus, standardizing domain terminologies is an essential activity in feature modeling. With the understanding of standard domain terminologies, we can clarify the scope of commonality and variability analyses and then identify common and different features within a product line efficiently and effectively.

Try to first find differences among products envisioned for a product line and then, with this understanding, identify commonalities. Products in the same product line share a high level of commonality. Hence, the "commonality space" would be larger to work with than the "difference space." It is our experience that finding differences is much easier than finding commonalities. We recommend, first, to identify product categories (e.g., freight and passenger elevators in an elevator product line), within each of which products should have a higher level of commonality and lower variability than those in other categories. Next, list features that characterize each category, i.e., differences between categories. Then, for each category, list products and do the same thing. With this understanding, we experienced that identification of commonalities became easier and we could proceed to listing common features and modeling them effectively and efficiently.

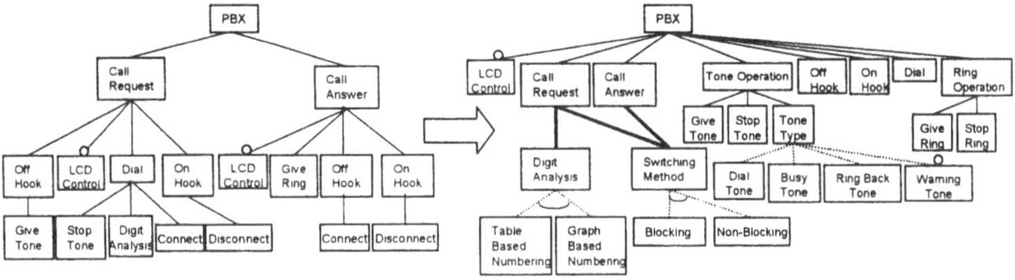

Figure 2.2 A feature organization example: a PBX[3] product line

Do not model implementation-oriented functions as features. Also, do not organize features to represent functional dependencies, like a function call hierarchy, but organize features to capture and represent commonalities and differences. Skillful developers tend to enumerate all the implementation details and identify them as features, even though there are no variations among them. It is important to note that a feature model is not a requirement model for representing internal functions and their behaviors. The modeling should focus on identifying properties, factors, and assumptions that can characterize products of a product line or that can differentiate one product from others in the same product line, not on finding all implementation details that are necessary to implement the products. Moreover, features should be organized so that commonalities and variabilities can be recognized easily rather than representing interactions among features, like a function call hierarchy. For example, the feature diagram on the left in Figure 2.2 merely enumerates all operations related to Call Request and Call Answer and organizes them like a functional call hierarchy. However, the feature diagram on the right shows common services (e.g., Call Request), operations (e.g., Tone Operation), and domain technologies (e.g., Switching Method) and also shows how they are different among products in the product line. For instance, Connect and Disconnect in the feature diagram on the left are grouped into a common domain method, i.e., Switching Method, which is further refined to Blocking and Non-Blocking to express variability in the product line.

Refine a feature if its refined "sub-features" are associated with other features, even though no variation exists among products in a product line. A feature needs not be refined further into "sub-features" if there are no variations among different products. However, even if there is no variation among different products for a feature, it may be

3 PBX (Private Branch Exchange) is a switching system, which is used within a relatively small area such as a hotel, an office building, or a hospital.

refined further if its refined "sub-features" are associated with different features. For example, in the elevator control software, Weight Detection Method is a domain specific method feature that is used to implement two operation features, Motor Control and Call Handling. However, this feature is used somewhat differently for different features: for example, Motor Control mainly uses Automatic Weight Detection and Call Handling uses Predefined Weight Detection; therefore, Weight Detection Method is refined further, as shown in Figure 2.3.

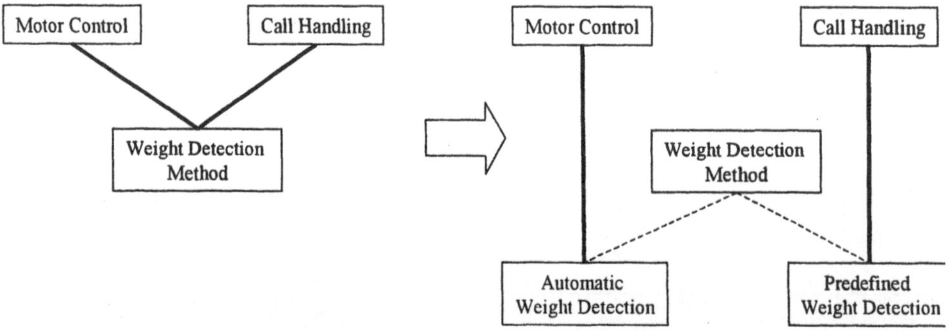

Figure 2.3 A feature refinement example: an elevator product line

Note that the focus of feature modeling is on identifying commonalities and variabilities among products in a product line and organizing them so that they can be understood easily and be useful in engineering component assets. FORM supports the design of product line architectures based on a feature model, and principles and guidelines that can be used during the design are discussed in the following section.

2.4 ARCHITECTURE DESIGN PRINCIPLES

There are several architecture design principles that have emerged from our experience of working on several industrial product line projects. Some of these principles are discussed in this section.

Separation of functional, process, deployment, and module architecture views. Separating different aspects of systems into separate views helps architects manage complexity of large-scale product line architecture design. Each of four architecture views addresses different engineering concerns during the architecture design and refinement. In the functional architecture view, cohesion of each component and coupling among architectural components are major engineering issues; performance, coordination and synchronization among processes are major concerns in the process architecture view; resource allocation and run time quality issues, such as performance and fault tolerance, are addressed in the deployment architecture view; and reusability and adaptability are the primary engineering concerns in the module architecture view[4]. By separating different engineering concerns in different architecture views, architects can analyze design tradeoffs of the product line architecture more easily.

Separation of components from connection mechanisms. In the development of product-line architectures, FORM supports postponement of decisions on component

4 The module architecture view represents a detailed design structure of architecture components, which is a final output of product line asset development.

connection mechanisms, which is depicted in Figure 2.4. In the functional architecture, information flows among architectural components are defined, which are refined into logical interaction models such as message queue, message-reply, etc. in the process and deployment architectures. These logical interaction models are implemented for various communication techniques that come with the operating system or middleware selected for a product. Binding of product line software with particular communication techniques can be made at the module instantiation time considering performance and architectural configurations (e.g., distributed vs. centralized architecture). This principle implies that selection of a particular connection mechanism does not impose any restriction on selection of architectural components (i.e., functional components, process components, and module components).

Functional Architecture	**Data Flow**			**Control Flow**
Process/ Deployment Architectures	Message Queue	Message/Reply	Shared Data	Event
Module Architecture	Server Socket, Client Socket, Datagram Socket	Procedure Call	Shared Memory	Implicit Invocation
Implementation Techniques	Local Socket, UNIX Socket, Java Socket	Local Procedure Call, ONC RPC, DCE RPC, CORBA, Java RMI	Local Shared Memory, UNIX Shared Memory, Remote Shared Data	Function Pointer (optionally with UNIX Socket or Java Socket)

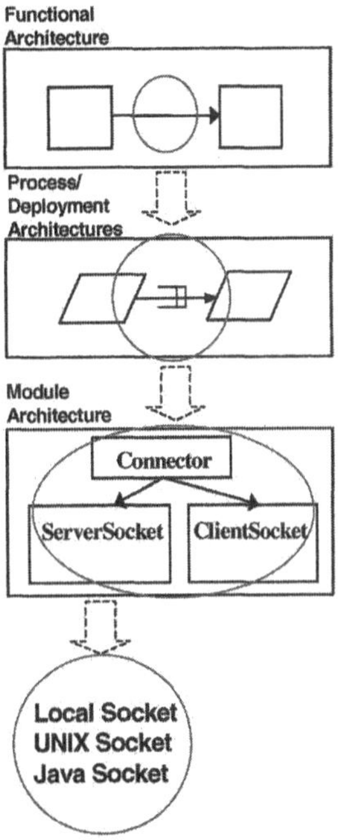

Figure 2.4 Architecture connection mechanisms

Correspondence between feature boundaries and architecture boundaries. Product-line software should be adaptable for a set of features selected for a specific application. By designing each selectable feature as a separate component, applications can be derived easily from product-line software. If there is difficulty in establishing this relation, the feature must be refined into specific features so that features can be easily mapped into architectural components. This will enhance the traceability between architectural components and features.

Parameterization of the product line architecture using non-functional features. There may be several architectures with the same functionality in a product line. Nonfunctional features, such as quality attribute and cost, are important factors that can differentiate product line architectures. During the architectural design and refinement, non-functional features are used to allocate their associated features to architectural components or to select components from alternative ones with the same functional features. For example, in the elevator control software product line, Motor Control Latency is an important performance feature, and the two features, Velocity Profile and Motor Device, are related with Motor Control Operation. In this case, we had allocated these two features in two different ways as shown in Figure 2.5. In the Architecture 1, the two features are allocated to the same architectural component (i.e., MotorController), but in the Architecture 2 they are allocated to two different architectural components (i.e., MotorController and VelocityGenerator). The Architecture 1 is designed for high functional cohesion between the two features, but the Architecture 2 is designed for high motor control performance by separating VelocityGenerator, which can be allocated to different hardware, from MotorController. To support these design alternatives, the non-functional feature (i.e., Motor Control Latency) was used to parameterize the product line architecture.

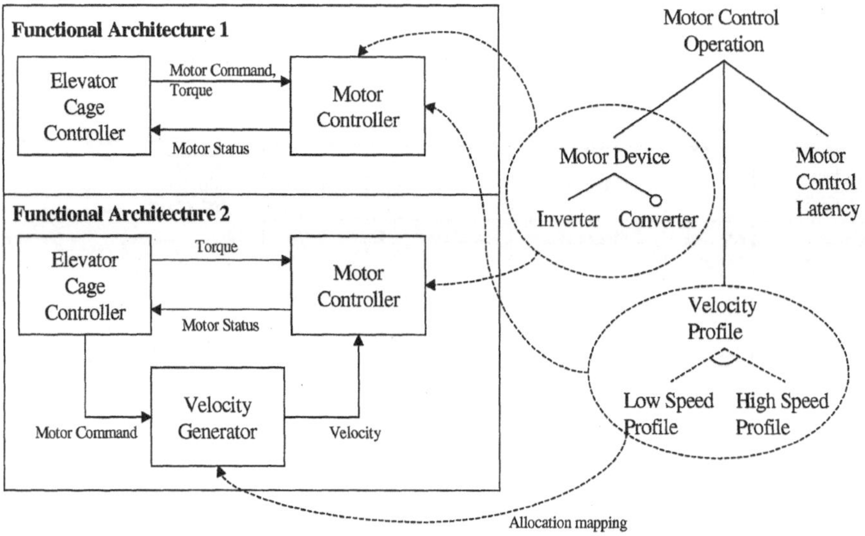

Figure 2.5 Parameterization of the product line architecture

Once architecture models are constructed, components identified in the models are refined with detailed information necessary for implementation. Reusability across these components must be considered during the refinement. Component design principles and guidelines are in the section below.

2.5 COMPONENT DESIGN PRINCIPLES AND GUIDELINES

There are engineering principles and guidelines that we can apply in the development of components for a product line. It should be noted that these principles and guidelines be applied with consideration of quality attributes required by the product line.

Separation of common features from product specific features. The core asset of a product line includes components implementing the common features of the product line and those that are frequently included in the products as utilities or as components for product specific features. These components are integrated, adapted, and extended to create specific products and, therefore, they should be designed with consideration of how they will be adapted to product specific features and other possible extensions. Adaptability and extensibility must be built into the core asset component design.

Encapsulation of product specific features (function, data, and control). Features specific to each product and, therefore, can vary from one product to other, should be separated from the components implementing the core features and be encapsulated in separate components with appropriate interfaces. These interfaces that encapsulate product specific contexts (e.g., computing environment, cultural feature, devices, etc.), functions, data, and control will allow adaptation and extension of these product-specific features, and integration with the core components (implementing the common features).

Postponement of context binding. Components should be designed so that binding with the product and/or customer specific contexts (e.g., computing environments, product specific features, etc.) may be postponed until it is needed. Context bindings may occur during product development, at installation time, and/or while the system is in operation. When and how the bindings should occur must be decided based on the marketing and product plan.

Separation of global control and local control for additive features. The product line asset components must be reusable and composable, and accommodate the variability among products through different composition configurations. For this to occur, controls that require the state information of the system (i.e., global controls) must be separated from controls that require only the local information (i.e., local controls). The separation and the mechanisms employed to implement this separation will allow analyses of the global control behaviors of a product when components implementing various features are added, removed, or modified. Selection of particular "separation mechanisms" must be made with consideration of the quality attributes required by the product line. In addition to the above principles, components can be made adaptable and modifiable by separating control, data transformation, persistent data management, and interface functions, and by separating components from component connection mechanisms. With this separation, the ripple effects caused by a change can be isolated to a small number of components.

We found that product line components are classified generally into the following types:

- Workflow management components: These components implement business processes in business applications, and control sequences, task synchronization and coordination in embedded control systems.
- Functional components: These components implement the core functionalities of a product line: for example, business functions (e.g., transactions) in business applications, and data transformations in embedded control systems.
- Domain technology components: These components encapsulate "enabling technologies" that are used to implement the core functionalities of a product line. For example, the technologies for transferring financial or credit information over the Internet are critical in e-commerce systems. These technologies can change independently from the functionalities of the application and therefore they should be separated from the functional components. Also, they may be used in many different applications.
- Interface components: These components encapsulate interface devices, protocols, data conversion (e.g., marshaling, un-marshaling), data filtering, and packing and unpacking, etc.
- Data sharing components: These components serve as data repositories and they encapsulate particular structures and the technologies or components that may have been used to implement the repositories.
- Communication/synchronization components: These components encapsulate particular mechanisms used for communication and synchronization among components. Typically, these components hide the middleware or OS primitives used to implement the communication and synchronization mechanisms.

This classification can serve as useful guidelines in identifying asset components of a product line.

2.6 PRODUCT LINE ENGINEERING: YARD INVENTORY SYSTEMS

In this section, we illustrate the product line engineering of the yard inventory systems of an iron and steel company in Korea. The company has more than twenty five "yards" to store and manage intermediate products between two plants. The yard inventory systems have typically been replaced by new ones about every seven years and they are modified about twice during their lifetime due to the changes of external environments (e.g., addition or removal of equipments or changes of system environments). That is, 3.6 systems are replaced and 7 systems are modified each year. Over the years, the development and maintenance costs of the systems have not been reduced, even though the company has developed or modified similar types of systems every year. This is mainly due to the fact the company does not have reusable core assets (i.e., standard architectures and reusable components). In order to improve the productivity and reduce the time to deployment, the company has decided to develop reusable assets for the yard inventory systems.

Details of the product line engineering are not included in this section. Instead, we illustrate how the guidelines and principles discussed earlier were applied to the product line engineering of yard inventory systems.

2.6.1 A Yard Inventory Management Product Line

A yard is a space for keeping steel products (e.g., slabs or coils) temporarily before transferring them to the next manufacturing processes or delivery to customers. Yard inventory management product line is a set of systems whose main function is to determine where and how to pile up products within each yard. In addition, the systems track the positions and states of products and control various equipments (e.g., cranes, conveyer belt, and transfer vehicles) to move them to different locations within a yard.

The primary services of a yard inventory management product line are to make work plans for crane operation considering the types and states of products, and the production and market plans. These services can be classified into three categories: storage, delivery, and transfer services. The storage service is to pile up products entered into a yard efficiently; the delivery service is to take them out of a yard; and the transfer service is to move them within a yard to rearrange for the next manufacturing process.

2.6.2 Feature Modeling

As stated in the first feature modeling guideline, feature identification starts with analyzing standard terminologies of the yard inventory management product line as shown in Table 1. As a first step, we identify product categories of a product line and then analyze commonalities and differences among different product categories. In the yard inventory management product line, there are two product categories, coil yard and slab yard. The commonality and difference between two product categories are shown in the right two columns of Table 2.1. The next step is to list products for each category and then analyze differences further. Table 2.2 shows the differences among products belonging to the slab yard category.

After identifying all common and different features within a product line, they should be organized so that commonalities and variabilities can be recognized easily. Figure 2.6 shows a simplified feature model, which captures composition, generalization, and implementation relationships among common and different features within the yard inventory management product line. The product line consists of two types of common service features, Work Planning and MMI. Work Planning is further specialized into three service features, Storage, Transfer, and Delivery. These service features share common operation features (e.g., Lift Control Planning, Moving Control Planning, Priority Based Scheduling, etc.). Of these features, Transfer, Moving Control Planning, and Priority Based Scheduling are modeled as optional features as these features may not be included in some products. (See Table 2.2.)

The Lift Control Planning Preparation feature is implemented by two domain technology features, Work Boundary Selection and Work Unit Determination. Work Unit Determination is modeled as an optional feature as this feature may not be required in a yard where cranes can lift only one slab at a time. Work Unit Determination can be implemented by Grouping By Crane Capability and Grouping By Code and, optionally, by Grouping By Sequence. Work Boundary Selection is implemented only by Grouping By Crane Capability.

Table 2.1 Standard terminologies of a yard inventory management product line

Domain Languages	Definitions	Slab Yard	Coil Yard
Work Plan	Operation plans of cranes, moving equipments (e.g., conveyer belt or vehicles) to provides the service (e.g., storage, transfer, or delivery service) of yard inventory systems.	O	O
Storage	A service of piling up the products entered from the outside of a yard	O	O
Transfer	A service of transferring products within a yard	O	O
Delivery	A service of taking products out of a yard	O	O
Lift Control Planning Preparation	An operation of making a work plan for crane operation	O	O
Lift Control Planning Distribution	An operation of distributing work plans to cranes	O	O
Priority Based Scheduling	Scheduling work plans based on their priorities	O	O
Work Boundary Selection	Determining work boundary of a crane	O	O
Work Unit Determination	Determining the number of products to be lifted	O	
Grouping By Crane Capability	Grouping products as a work unit according to the crane capability	O	
Grouping By Code	Grouping products as a work unit according to the product codes	O	
Grouping By Sequence	Grouping products as a work unit according to the predefined work sequence	O	

Note that, as stated in the third feature modeling guideline, a feature model does not represent all implementation details that are necessary to implement the products. For example, in order to implement Work Unit Determination, some database manipulation functions are needed. But we need not identify all of these functions as features if they do not represent any significant differences among products in the yard inventory management product line.

Table 2.2 Commonality and variability analysis among different products

features / Products	Transfer	Moving Control Planning	Transfer Car	(Crane) Capacity	Work Unit Determination	Grouping By Sequence	Priority Based Scheduling
Slab Yard A	O	O		One or more	O		O
Slab Yard B	O	O	O	One or more	O	O	O
Slab Yard C				One			

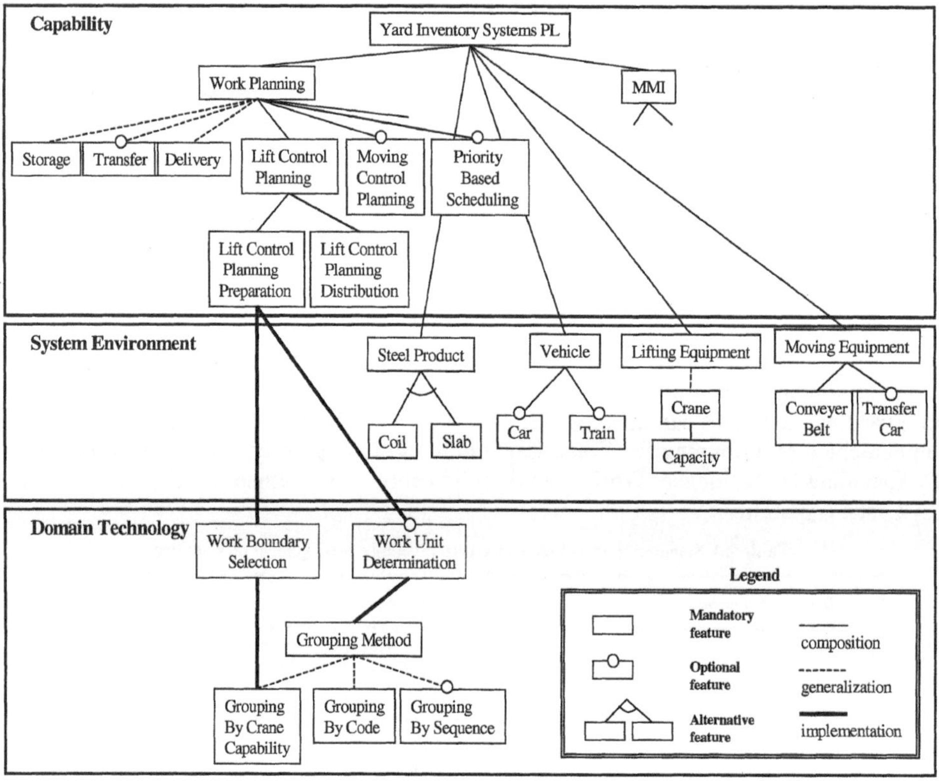

Figure 2.6 The feature model of the yard inventory management product line

2.6.3 Architecture Design

The major activity of architecture modeling is to allocate functional features to architectural components[5] (e.g., processes) with the considerations of non-functional requirements such as performance, an organization's policies, etc.

For example, Figure 2.7 shows the two alternative architecture models of the yard inventory management product line. The Architecture Model 1 in Figure 2.7 is designed to increase the performance of the systems by ignoring the factors (e.g., number of orders, urgent job request, states of all entities in a yard, etc.) required for optimal job scheduling. That is, CommunicationManager in the Architecture Model 1 distributes jobs to corresponding processes and each process makes a local schedule concurrently. On the other hand, the JobScheduler process in the Architecture Model 2 takes all factors into consideration and provides optimal schedules to the job control processes (i.e., VehicleControlPlanning, LiftControlPlanning, and MovingControlPlanning). The choice between the two architecture models is related to the selection of the non-functional feature Priority Based Scheduling in the feature model. (See Figure 2.6.)

5 In order to distinguish the component name from the feature name, we don't leave any space for component names, while putting spaces between words for feature names.

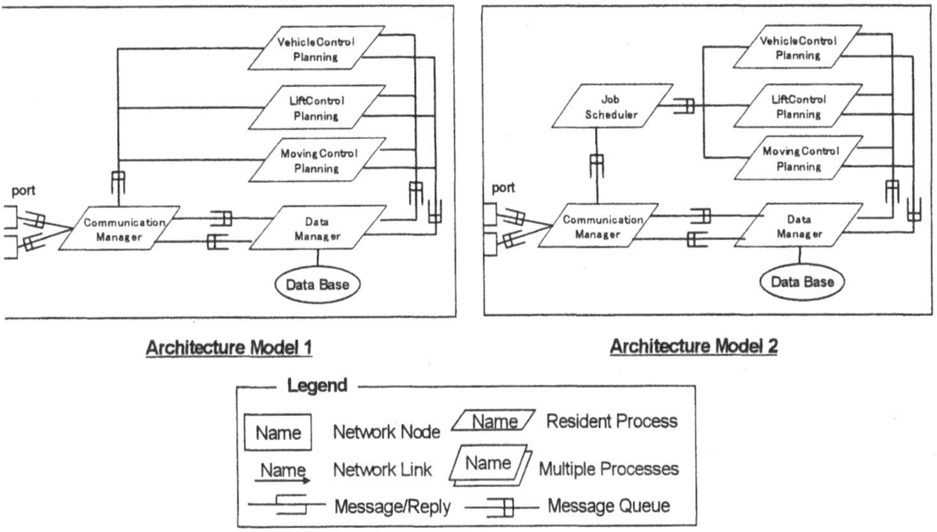

Figure 2.7 Alternative architecture models of the yard inventory management product line

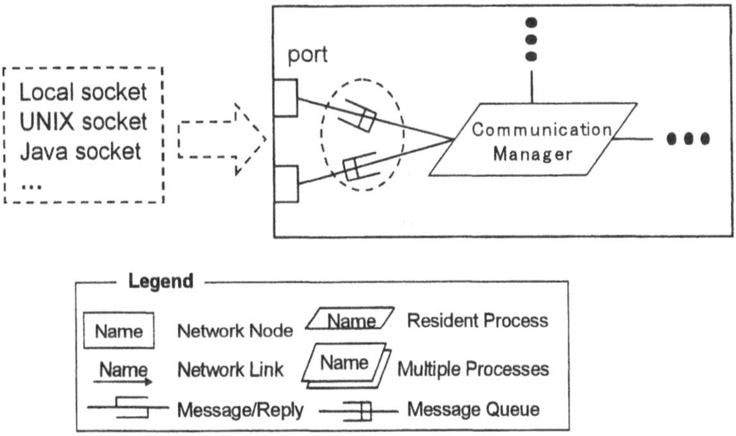

Figure 2.8 Various implementation methods for a message queue

The separation of connection mechanisms from architectural components allows easy adaptation to various implementation mechanisms. For this, our method provides various types of abstract "connectors" for different abstractions of architectures. (See Figure 2.4.) For instance, the communication mechanisms between the CommunicationManager process and other processes in external nodes are specified with Message Queue interfaces. During application instantiation, one might choose a particular data communication implementation of a Message Queue from various implementation mechanisms (e.g., local socket, UNIX socket, Java socket, etc.). (See Figure 2.8.)

During architectural design, a feature model may be refined when the physical boundary created by the architecture model does not correspond to the logical boundary created by the feature model. For example, the Lift Control Planning feature was originally allocated to a single process, LiftControlPlanning. However, the process was divided into LiftControl PlanningPreparation and LiftControlPlanningDistribution to meet the performance requirement for yard inventory systems with many cranes. To enhance the traceability between architectural components and their features, we refined the Lift Control Planning feature into Lift Control Planning Preparation and Lift Control Planning Distribution features, as shown in Figure 2.6.

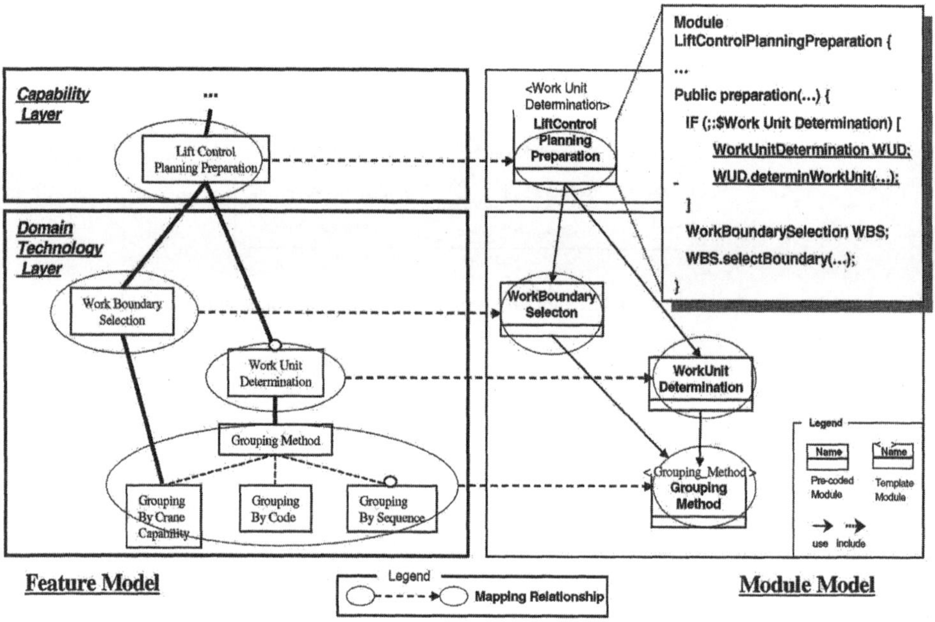

Figure 2.9 Mapping relationships between the feature model and the module model

2.6.4 Component Design

The purpose of component design is to produce components that can be easily reused and adapted in a variety of contexts. Component development starts with identifying objects potentially useful for the implementation of reusable and adaptable components. In our method, the feature model is used to derive components that encapsulate specific decisions (i.e., design decisions, implementation decisions, environmental constraints, etc.) that are made during product development.

Figure 2.9 shows how common features within the product line are implemented as core components and how product-specific features are encapsulated in separate components. For example, a common feature (i.e., Lift Control Planning Preparation) and a product specific feature (i.e., Work Unit Determination) are designed into separate modules (i.e., LiftControlPlanningPreparation and WorkUnitDetermination). This separation with appropriate interfaces makes it easy to adapt the module (i.e., WorkUnitDetermination), which encapsulates the product-specific feature. (See the mapping relationships in Figure 2.9.)

For context bindings to occur during product instantiation time, FORM macro processing is used to instantiate all parameterized (i.e., template) components. (See the upper right portion of Figure 2.9.) For instance, when the optional feature Work Unit Determination is selected, the macro processor leaves the underlined code segment in the LiftControl PlanningPreparation module to use the WorkUnitDetermination module. Otherwise, the LiftControlPlanningPreparation component excludes the code segment and uses only the WorkBoundarySelection module.

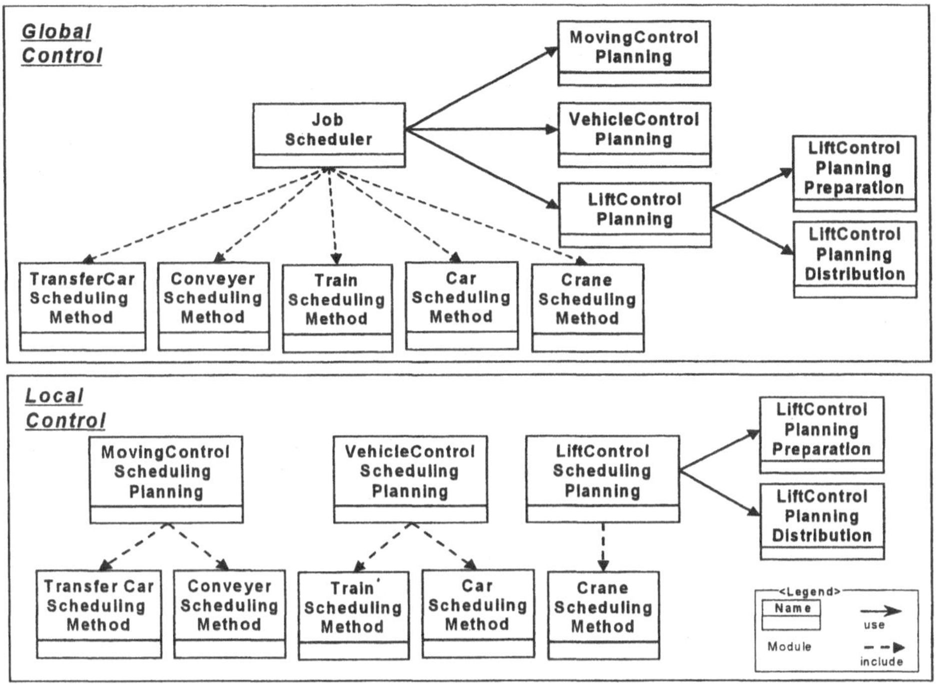

Figure 2.10 Separation of local control from global control for additive features

Figure 2.10 shows two design alternatives for job scheduling. The first design has the global control for job scheduling. That is, JobScheduler schedules jobs of MovingControlPlanning, VehicleControlPlanning, and LiftControlPlanning, considering all scheduling criteria of the entities in a yard such as car, crane, train, etc. In the second design, the control for scheduling is localized and distributed to MovingControl SchedulingPlanning, LiftControlSchedulingPlanning, and VehicleControlScheduling Planning. The second design can be selected over the first one as (1) local scheduling may be performed independently without considering the scheduling criteria of other planning components, (2) each scheduling component hides specific scheduling criteria locally, and (3) it is easier to add new planning components compared to the first design.

The following section concludes this paper with lessons learned from applying the FORM method to industrial product lines.

2.7 CONCLUSION

PLSE is one of the most promising approaches to increasing productivity and reducing the time to market by developing and evolving products from core assets of a product line. To develop reusable core assets, commonality and variability of a product line must be analyzed first and then the analysis results must be used to guide the development of core assets.

This paper has introduced a feature-oriented approach to core asset development for a product line with a focus on engineering principles and guidelines on feature modeling, architecture design, and component development. To demonstrate how these principles and guidelines are used in product line asset development, we applied them to the product line engineering of yard inventory systems.

FORM has been applied to several industrial product line projects including electronic bulletin board systems, PBX systems, elevator control systems, and yard inventory systems. Lessons learned from these projects are summarized as follows:

- Product line engineering, in general, requires a large initial investment and a long development time compared to single product development. When an organization does not have a product line experience, we recommend that a small product line, or a subsystem of a product line, that has a high level of commonality be selected for experimentation and experience. Also, it is imperative that the organization has in-house expertise with the selected product line and that there is no products waiting to use the output from the exploratory product line. After learning and exploring with a small product line, organizations may extend their experiences to large-scale product lines. We have observed that organizations starting with ambitious goals and expectations often fail due to high initial costs and technical difficulties.
- When a product line includes products with fixed delivery schedules, the manager tends to assign a higher priority to the development of the products than performing a detailed product line analysis as the time to deliver the products draws near. We recommend not to select a product line with products having fixed delivery schedules.
- We have experienced that a large degree of contextual differences (e.g., external devices that products interface with) between products envisioned for a product line has often been an indication of different sub-product lines within the product line. An understanding of the extent of the contextual differences among the products will help to re-scope the product line. For example, a freight elevator product line consisted of several types of products, for example, parking building elevators and warehouse elevators. The hardware devices each type of elevators interfaces with were vastly different and we found it very difficult to abstract out common hardware interfaces. It was much easier to handle these as separate product lines.
- We found that organizations with a standard system infrastructure (e.g., physical communication media) or a software platform were usually unwilling to accept the possibilities of changes to these standards. Engineers who had been building products using the standards tended to believe that their standards "work" and there would not be a need for a change. It is a job of product line analysts to convince them about the needs to prepare for potential changes as well as current variations, since any standards will have to evolve as technologies evolve.

Product line engineering requires a large initial investment and, often, structural and cultural changes of an organization are required. In order to succeed, a strong long-term commitment by the management is the key element.

ACKNOWLEDGEMENT

We would like to thank the Ministry of Education of Korea for its financial support toward the Electrical and Computer Engineering Division at POSTECH through its BK 21 Program. This work was partially sponsored by Korea Industry Technology Association through the project of commercialising ASADAL CASE tools.

BIBLIOGRAPHY

Cohen, S. G., Stanley Jr., J. L., Peterson, A. S. and Krut Jr., R. W., 1991, Application of Feature-Oriented Domain Analysis to the Army Movement Control Domain, in *Technical Report, CMU/SEI-91-TR-28*, ADA 256590, (Pittsburgh, PA: Software Engineering Institute, Carnegie Mellon University).

Czarnecki, K. and Eisenecker, U., 2000, *Generative Programming: Methods, Tools, and Applications*, (New York: Addison-Wesley).

Griss, M. L., 2000, Implementing Product-Line Features by Composing Aspects, In *Proceedings of the First Software Product Line Conference (SPLC)*, Denver, Colorado, USA, edited by Donohoe, P., *Software Product Lines: Experience and Research Directions*, (Norwell, Massachusetts: Kluwer Academic Publishers), pp. 271-288.

Griss, M. L., Favaro, J. and d'Alessandro, M., 1998, Integrating Feature Modeling with the RSEB, In *Proceedings of the Fifth International Conference on Software Reuse*, (Victoria, BC, Canada), pp. 76-85.

Hein, A., Schlick, M. and Vinga-Martins, R., 2000, Applying Feature Models in Industrial Settings, In *Proceedings of the First Software Product Line Conference (SPLC)*, Denver, Colorado, USA, edited by Donohoe, P., *Software Product Lines: Experience and Research Directions*, (Norwell, Massachusetts: Kluwer Academic Publishers), pp. 47-70.

Kang, K., Cohen, S., Hess, J., Nowak, W. and Peterson, S., 1990, Feature-Oriented Domain Analysis (FODA) Feasibility Study, *Technical Report CMU/SEI-90-TR-21*, (Pittsburgh, PA: Software Engineering Institute, Carnegie Mellon University).

Kang, K., Kim, S., Lee, J., Kim, K., Shin E. and Huh, 1998, M., FORM: A Feature-Oriented Reuse Method with Domain-Specific Reference Architectures, *Annals of Software Engineering*, **5**, pp. 143-168.

Kang, K., Kim, S., Lee, J. and Lee, K., 1999, Feature-Oriented Engineering of PBX Software for Adaptability and Reusability, *Software-Practice and Experience*, **29**(10), pp. 875-896.

Krut, R. and Zalman, 1996, N., Domain Analysis Workshop Report for the Automated Prompt Response System Domain, in *Special Report, CMU/SEI-96-SR-001*, (Pittsburgh, PA: Software Engineering Institute, Carnegie Mellon University).

Lee, K., Kang, K., Chae, W. and Choi, B., 2000a, Feature-Based Approach to Object-Oriented Engineering of Applications for Reuse, *Software-Practice and Experience*, **30**(9), pp. 1025-1046.

Lee, K., Kang, K., Koh, E., Chae, W., Kim, B. and Choi, B., 2000b, Domain-Oriented Engineering of Elevator Control Software: A Product Line Practice, In *Proceedings of the First Software Product Line Conference (SPLC), Denver, Colorado, USA*, edited by Donohoe P., *Software Product Lines: Experience and Research Directions*, (Norwell, Massachusetts: Kluwer Academic Publishers) pp. 3-22.

Simos, M., et al, 1996, Software Technology for Adaptable Reliable Systems (STARS) Organization Domain Modeling (ODM) Guidebook Version 2.0, *STARS-VC-A025/001/00*, (Manassas, VA: Lockheed Martin Tactical Defense Systems).

Vici, A. D. and Argentieri, N., 1998, FODAcom: An Experience with Domain Analysis in the Italian Telecom Industry, In *Proceedings of the Fifth International Conference on Software Reuse*, (Victoria, BC, Canada), pp. 166-175.

Zalman, N. S., 1996, Making the Method Fit: An Industrial Experience in Adopting FODA, In *Proceedings of the Fourth International Conference on Software Reuse*, (Orlando, FL), pp. 233-235.

CHAPTER 3

Software Architecture for Intelligent CAD Systems

Toyohiko Hirota

Kyushu Sangyo University

Masa-aki Hashimoto

Kyushu Institute of Technology

ABSTRACT

This paper proposes software architecture for intelligent CAD systems, and describes the implementation of a prototype. The problems encountered in using CAD systems in industry involve complicated data handling and unsatisfied demands for domain knowledge arising due to the lack of a method by which to extract it and adopt it to the system. In the sample domain of the architecture, we have already defined domain-specific BDL (Building Design Language), for use by architecture experts in describing modelers of architectural structure in CAD systems. In addition, we have developed a CAD system generator based on BDL descriptions. However, the different domain-specific languages required for individual domains make it difficult to develop various CAD system generators. The proposed software architecture solves this problem by applying a common intermediate language based on the ER model. Moreover, it allows the creation of an integrated CAD system containing multiple domains required by various fields of industry. The implementation of its prototype demonstrates its feasibility.

3.1 INTRODUCTION

Almost all industries in advanced countries have been using CAD (Computer Aided Design) systems for more than two decades. Most companies have installed some kind of CAD workstation and software package to facilitate the special tasks involved in the design process. However, in the course of fulfillment of an engineering commission, the penetration of using CAD systems is lower than 20% (Abeln, 1997). Major problems in operating CAD systems in industrial setting arise due to the difficulty of product data exchange between various CAD systems, and from conventional problems such as complicated data handling and unsatisfied demands resulting from the method by which to extract domain knowledge and adopt it to the system. The former problem is due to the diversity of data formats and standards, which gives rise to conversion problems (De Martino, *et al.*, 1998).

We have been working to solve the latter problem by studying an efficient method for developing reliable and highly quality application software. To develop high quality software applied to a domain, the domain experts themselves should describe the requirements. For this purpose, a specification language comprehensible to the experts should be defined. This language must be specific to the domain so as to enable experts to describe the system requirements on the basis of their particular skills. Moreover, the descriptions should be automatically transformed into programs in order to ensure development efficiency and software reliability.

In the sample domain of the architecture, we have already defined a domain-specific BDL (Building Design Language) (Hirota, *et al.*, 1995) for use in describing the modelers of architectural structure in CAD systems. Moreover, we have developed a CAD system generator based on BDL descriptions. A function for designing architectural structure and fire alarm systems has been implemented. We have also been working to develop languages for describing the modelers of space arrangement in a building (Hashimoto, *et al.*, 2000), and of building construction scheduling and management (Shin-yashiki, *et al.*, 1998) (Miura, *et al.*, 2000) (Aman, *et al.*, 2000).

Those languages should be able to express the significant qualities of objects and abstract structures, such as existence dependency, behind the whole objects in the domains. Therefore, those languages have adopted domain-specific models obtained from the ER model by imposing the restrictions associated with the abstract structures. However, different domain-specific models are obtained from different individual domains. It would be difficult, not to mention impracticable, to develop a variety of CAD system generators, one for each individual domain.

In order to develop the CAD system generators efficiently, we propose software architecture for intelligent CAD systems, and describe the implementation of its prototype herein. The principle idea of our software architecture is to utilize domain-specific models obtained from the ER model by imposing restrictions, and to generate a CAD system automatically through the ER model. Section 3.2 describes the domain-specific conceptual model. Section 3.3 and 3.4 discuss general data models and computation methods for CAD systems. In section 3.5, our software architecture is applied to CAD system generation. Section 3.6 describes its prototype implementation. Section 3.7 summarizes our work.

3.2 DOMAIN-SPECIFIC CONCEPTUAL MODEL AND CAD SYSTEM DEVELOPMENT

There are a large variety of tools referred to as CAD (Computer Aided Design). The most general and widely used of these are drawing tools used to prepare plans. However, most drawing tools do not directly support design. When such a drawing tool is used to prepare a plan, the user concentrates his attention on the operation of the tool, and it is difficult to focus simultaneously on the plan itself. Therefore, the designer drafts the plan by some other means, and a CAD operator subsequently makes a fair copy of the plan by using a CAD tool. In such a situation, however, information technology is useless to the designer's task.

On the other hand, in many enterprises, tools to design a specific kind of product or part automatically have been developed and used. These tools are informed by the knowledge of the experts, which is formalized to be embedded in the tools. By using such tools, even a novice designer can execute a draft of a plan on an average level.

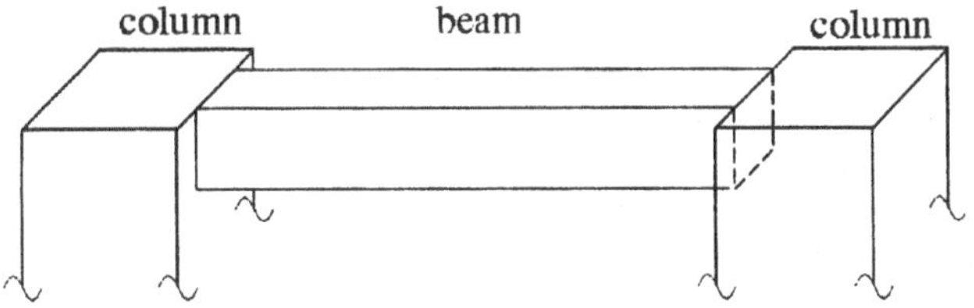

Figure 3.1 Connection of Parts

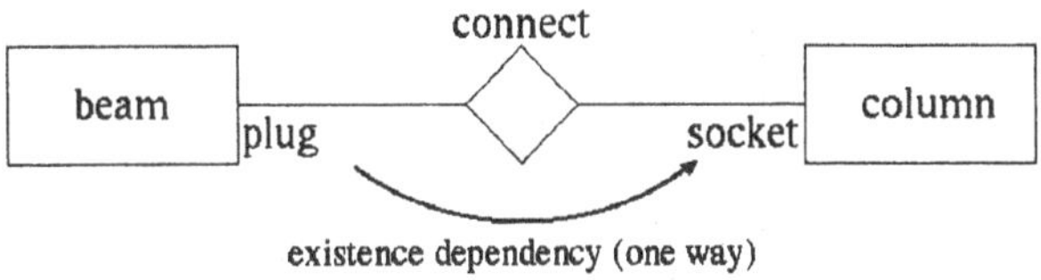

Figure 3.2 ER model of parts

Typically, however, these tools are difficult to maintain after they have been established. It is not easy to modify tools or adapt them to changes in the environment, new technology, or changes in design method. This is because

(1) design knowledge is not separated from programming technique, and
(2) description of design knowledge is not adequate.

To solve the former problem, we prepare a framework within which design knowledge is described. The second problem is that only programming experts can read the descriptions. Because they are usually not design experts, programming experts cannot grasp the contents of the description. The design knowledge descriptions are likely to be incomprehensible to anyone except the person who wrote them.

One approach by which to avoid the above-mentioned problems involves domain analysis and modeling (Itoh, *et al.*, 1998). Domain analysis and modeling has been adopted as a means for promoting software reuse. It is also effective in putting design knowledge for CAD in order. In this case, the target domain is the field of design, or the design knowledge of the experts in the field. We analyze the domain, and construct a domain model with which we can represent the domain knowledge. For example, a specification language, a diagram, or a table can be a domain model. A domain model should

(1) be comprehensible to the designers, and
(2) have a proper formal model as its background.

We previously analyzed the building design domain (Hirota, 1998). In the structural design of a building,

(a) a building is a set of parts, and
(b) a part is located relative to reference lines or other parts.

Figure 3.1 shows that a beam should be connected to a column at either end. This relationship can be represented by the ER model shown in Figure 3.2. The relationship between parts, however, is not a general one because it has a direction of existence dependency. For example, a beam must be connected to the columns. If it does not connect to the columns, the beam cannot exist. The connection is a necessary condition of the beam. On the other hand, the columns can be connected to beams, but the connection is not a necessary condition of the existence of a column. Therefore we use "plug and socket" as a metaphor for the relationship. We model a girder as a part with two plugs, and a column as a part with four sockets. Each plug of a beam must be connected to a socket of a column.

A part has many attributes, which are classified into two categories. Attributes in the first category are independent of other parts. For example, the width and height of a girder can be determined independent of the connected columns. Attributes in the second category are dependent on those of the parts that are connected via the plug and socket. For example, the length of a beam depends on the positions of the connected columns. Unlike existence dependency, attribute dependency is not one-way. The load of a column depends on those of the connected beams.

Based on the above discussion, we have defined, as a domain model, a domain-specific specification language BDL (Building Design Language) described in 3.6.1. We have written various kinds of specifications with BDL to ascertain its usefulness and that of our CAD generator (Hirota, 1998) (Nakashima, *et al.*, 2000). Moreover, we have applied our approach to architectural design (Hashimoto, *et al.*, 2000) and equipment installation design (Nakashima, *et al.*, 2000). In these studies, the following new problems of CAD system development have come to the surface:

(1) efforts to develop many similar generators,
(2) data exchange between different CAD tools, and
(3) various computation methods.

We developed BDL specific to the structural design of building. This BDL is not suitable for architectural design, equipment design, and so on. There are two approaches to solving this problem. The first is to extend BDL so as to specify architectural design or equipment design. In that case, BDL would become a more general specification language. This approach, however, reduces the advantage derived from the specificity of BDL. The second approach is to develop a new language for each domain, in which case one generator is developed for each language. To reduce the work of developing many generators, we adopt an intermediate model. Specifications written with a certain language are translated into the intermediate model, then translated into a programming language. The latter is a translator common to all specification languages. In section 3.3, we discuss several data models as the intermediate model for the translation.

We aim to develop an integrated CAD system. However, a wide variety of CAD systems are in practical use. Therefore, our CAD systems should be able to exchange CAD data with other existent CAD systems. To implement the data exchange, we must write a conversion program for every combination of CAD systems. The above-mentioned intermediate model is also useful to reduce the work of writing a conversion program.

Our specification languages are nonprocedural: they are superior in terms of minimality, understandability, extensibility, etc., to procedural languages (Hirota, *et al.*, 1995). Subsequently we must add computation methods to the specifications when we generate a CAD system. However, there are several kinds of computation methods. The method is selected based on the component in use. In section 3.4, we discuss several computation methods for CAD systems.

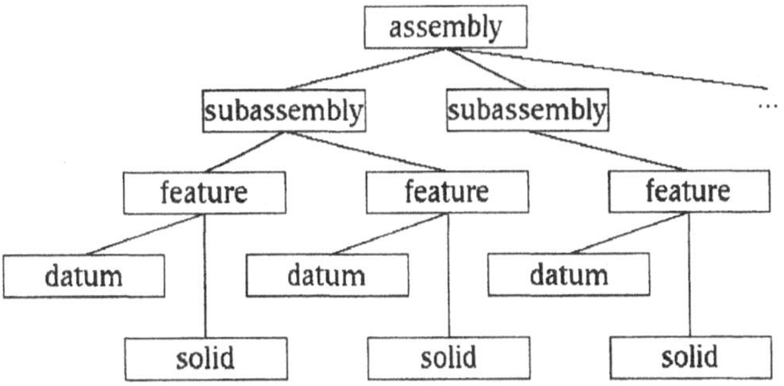

Figure 3.3 Model structure of Pro/ENGINEER

3.3 DATA MODEL FOR CAD SYSTEM

Though there are many kinds of data models, we focus on the following four, which are suitable for CAD:

(1) tree structure model,
(2) relational model,
(3) ER model, and
(4) object model.

3.3.1 Tree Structure Model

The tree structure model is suitable for representing the assembly structure of objects. Figure 3.3 shows a model structure of Pro/ENGINEER (PTC). The primary element of Pro/ENGINEER is a feature, and an object is an assembly of features. A feature consists of a datum and a solid. A datum is a plane, an axis, a point, a coordinate system, or the like, and is used for the reference of model construction. A solid represents a shape of model: that is, a cut, a protuberance, a hole, an angle, a surfacing, or the like. A cut or a protuberance is a 3-dimensional feature that is pressed out, rotated, swept, or blended from a 2-dimensional sectional sketch.

In Pro/ENGINEER, a user creates a complicated shape step-by-step, based on features with a simple shape. First the user creates a simple feature, then combines several features to make a part. Combined parts constitute a sub-assembly, and a product is completed when final assembly is achieved. There is an "a-part-of" relationship between an assembly and its parts. If an assembly is modified, its parts are automatically modified so as to reflect the modification of the assembly.

Pro/ENGINEER identifies two types of relationships, implicit relationships and explicit relationships. Implicit relationships are determined by assembly structure. For example, if a cube and a hole are combined, they are connected by a "has-a-hole-of" relationship. Explicit relationships express constraint relationships among feature dimensions, and to be described as arithmetic expressions by the user.

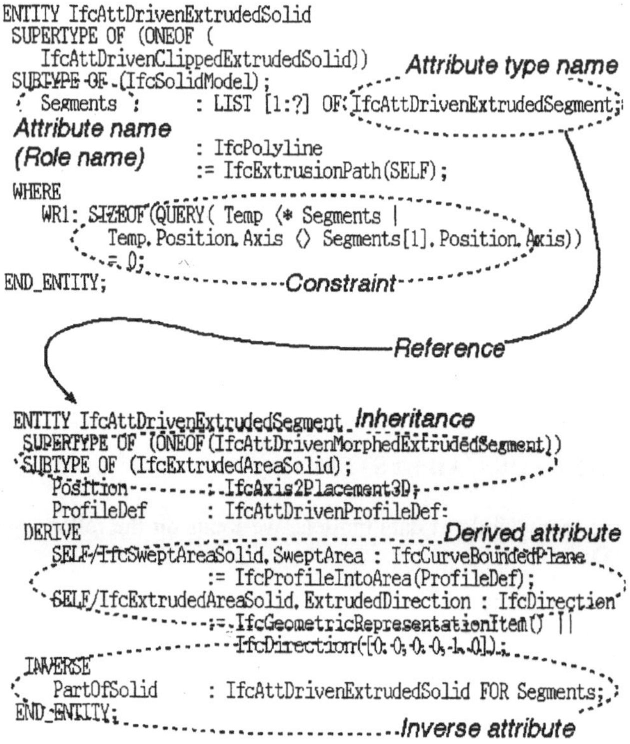

Figure 3.4 Entity type

3.3.2 Relational Model

The relational model was proposed by E. F. Codd for database schema (Codd, 1970). In the relational model, all data is represented as a set of tuples, and each tuple has several attributes. The relationship between two entities is also represented as a tuple: that is, as the combination of primary keys of the entities. IFC (Industry Foundation Classes) (International Alliance for Interoperability, 1997), the standard data model for CAD in the architectural industry, is described using the EXPRESS language of STEP (Fowler, 1995). The data model applied to EXPRESS language is based on the relational model, and contains the characteristics of the object-oriented data models, which are described in the following:

(1) Entity
An entity corresponds to an object. Therefore, an entity type chiefly describes a class of the IFC, as shown in Figure 3.4. Attributes, class inheritance, and class constraints are defined within the entity type.

(2) Attribute
The attribute values of an entity determine its properties. A type of attribute value can be REAL or BOOLEAN. SET, LIST, ARRAY, and BAG are also types of attribute values. OPTION means that the attribute may have no value.

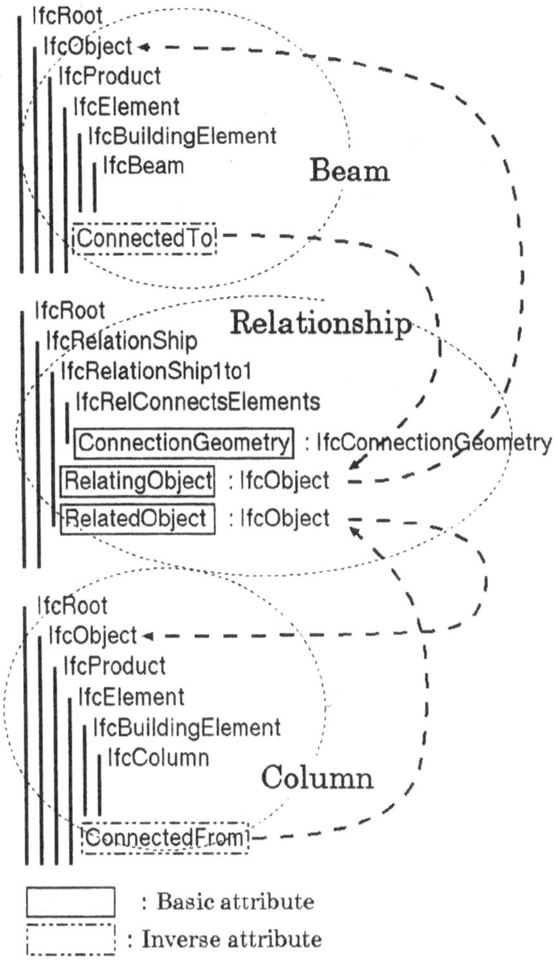

Figure 3.5 Example of relationship entity

The attributes of the IFC entities are classified as basic attributes, derived attributes, or inverse attributes. Basic attributes have values that are determined independent of each other. On the other hand, derived attributes obtain their values from other attribute values via calculations such as FUNCTION and PROCEDURE definitions.

Attributes can also have entities as their value types. That is, the type descriptions define binary relationships between pairs of entities. Strictly speaking, they determine one-way mappings from entity to entity. The names of such attributes specify the roles of entities in the relationships. When two-way mappings are necessary, the above-mentioned inverse attributes are applied.

(3) Relationship Entity

By using only the above-mentioned attributes that refers to entities, an entity can be made to express a relationship between entities, as shown in Figure 3.5. In the figure, class 'IfcRelationship' determines relationships between beams as relating objects and columns as the objects to which they are related. The inverse attributes 'ConnectedTo' and 'Connected-From' are applied to both beams and columns.

```
INFORMATION
  E column
    EN -100
   EC .col_is_created..created_col_cmd
   A old_xdiam NUM
      = NIL FROM col_is_created
      = NIL FROM col_is_deleted
   A new_xdiam NUM
      = ...
  ...
  R west_xbeam_is_connected
   C west_xbeam.column
     RN -1
   C east_col.xbeam
     RN -1
   RC (east_col.xbeam:
        created.xbeam_is_created.cmd.create_xbeam_cmd.w_x_sn
        == west_xbeam.column:
        xbase.xbase_is_connected.col.cline.new_sn)
     && ...
```

Figure 3.6 Example of PSDL descriptions

(4) Inheritance
For example, class 'IfcColumn' inherits its property from its super classes defined in the five levels. Thus, the inheritance hierarchy of the IFC is composed of many levels.

(5) Algorithm
FUNCTION and PROCEDURE statements describe algorithms for calculating attribute values. WHERE statements in entity descriptions define restrictions on the attribute values of their own entities and other entities.

3.3.3 ER model

The ER model was proposed by P. P. Chen as a unified model for a database (Chen, 1976). The ER model represents a target domain by entities and relationships. Because the ER model is more straightforward than the relational model, it is used for the analysis and documentation in designing relational schemas.

PSDL (Hashimoto and Okamoto, 1990) is a non-procedural specification language based on the ER model. It is mainly used for business processing, and the PSDL compiler can generate a C program from PSDL specifications. We have written, in PSDL, building parts and their constraints, as shown in Figure 3.1. Figure 3.6 shows an example of PSDL description (Hirota, *et al.*, 1995). The description begins with the reserved word INFORMATION[1].

(1) Entity type and attribute
Each entity type is described by the E (Entity) statement. In Figure 3.6, the entity type 'column' is described. Attributes are described by the A (Attribute) statement. The two attributes 'old_xdiam' and 'new_xdiam' are described in Figure 3.6. In PSDL, a variable such as an attribute value can be only assigned once, and it cannot be modified.

[1] The other words are DATA and ACCESS, neither of which was used in this experiment.

Therefore, we need an old variable and a new variable for each substantial attribute in order to describe an operation that modifies the attribute.

(2) Relationship type
Each relationship type is described by the R (Relationship) statement. Following this statement, the entity types related to each other by the relationship type are described by the C (Collection) statements. The entity types and their roles are specified in the form 'role.entityType'. Following a C statement indicating an entity type, the number of relation-ships related to one entity in the entity type is described by the RN (Relationship Number) statement. In Figure 3.6, the relationship type 'west_xbeam_is_connected' is described, and it connects 'column' in the role of 'west_xbeam' and 'xbeam' in the role of 'east_col'. 'RN –1' means that the relationship number is 0 or 1.

(3) Attribute value Dependency constraint (AD)
An AD is described by the = (equal) statement. The attributes referred to by AD are described in the form 'role1.relationshipType.role2.entityType.attribute', which refers to the attributes of other entities via relationships. Role 1 is the role of the entity whose attribute value is obtained, and role 2 is the role of other entities.

(4) Relationship existence Dependency constraint (RD)
In the succession of R and C statements, the condition expression for specifying the existence of the relationship is described by the RC (Relationship existence Condition) statement. In Figure 3.6, The RC of 'west_xbeam_is_connected' is described. This condition means that x-, y-, and z-coordinates of the connection points of the parts are equal.

(5) Entity existence Dependency constraint (ED)
This constraint is used to obtain the new entities in an entity type. ED is not used in Figure 3.6.

3.3.4 Object model

The object model is the most popular model for specifying the static structure of a program (Rumbaugh, *et al.*, 1991). It has many convenient mechanisms by which to abstract concepts. For example, inheritance makes it easy for us to reuse source code, and polymorphism enables us to simplify program control. These mechanisms are useful especially for GUI programs. However, they are not so important for representing data structures. The IFC described in 3.2 adopts the inheritance mechanism, which makes each entity less comprehensive by contrast.

We have been developing a GUI generator for CAD tools (Miyazaki, *et al.*, 1999). A user has only to construct a class diagram on the screen of our GUI generator, then the generator can automatically generate a GUI subsystem. Each class in the diagram is a predefined one or a newly defined one for which the user writes the source code. Even when the user defines a new class, the new class typically inherits a certain predefined class, which reduces the work of writing source code.

3.4 COMPUTATION METHOD FOR CAD SYSTEM

We classify computation methods from the viewpoint of the sequence of computation:

(1) procedure-driven,
(2) event-driven,

(3) data-driven, and
(4) request-driven.

In procedure-driven computation, the computation proceeds according to a predefined procedure. Most programs written in any procedural programming language belong to this type. Other types of programs must be compiled into machine language programs whose type is procedure-driven.

In event-driven computation, an event that occurs randomly initiates a computation that is bound to the event in advance. GUI programs belong this type. Various events such as pushing a button, selecting an item from a menu, inputting characters in a field, or clicking a mouse initiate a procedure bound to the event, which may read data, update a screen, start a certain computation, and so on.

Data-driven computation is initiated when data necessary to the computation are prepared. An expression to compute necessary data will be computed when every piece of data in the expression has been obtained. When specifying data-driven computations, a ‘single assignment’ rule is usually used. Therefore, the relation among multiple computations is automatically extracted. For example, an expression to compute variable ‘x’ must precede other expressions that refer to the variable ‘x’. Because the ‘single assignment’ rule prohibits a loop, some special statements to describe a repetition might be introduced. In JSP (Jackson’s Structured Programming) (Jackson , 1975), the repetition of computations is derived from the structures of input and output data.

In request-driven computation, an expression to compute data is initiated when the data is requested. The sequence of computation is in the reverse order of that of data-driven computation. A part is displayed on screen, and there are numerous ways in which it can be displayed. Consequently, it is wasteful to prepare all necessary data in advance. The computation should be initiated according to how the part will be displayed. In this case, request-driven computation will omit useless computation.

When a variable is updated, the procedure-driven method and the data-driven method will recalculate the variables dependent on the updated variable. However, in the event-driven method and the request-driven method, there are no means to initiate the recalculation. One of the solutions to this problem is to recalculate all necessary variables when a variable is accessed. However, that wastes processing time. TMS (Truth Maintenance System) manages the dependency among variables to avoid this waste (Doyle , 1979). When a variable is updated, all the variables dependent on the update variable become void. Then the computation method recalculates any voided variable. For example, the CPU time of several editing operations in our CAD was reduced to 0.18 ~ 0.08 by using TMS (Hirota, *et al.*, 1998).

3.5 SOFTWARE ARCHITECTURE FOR CAD SYSTEM GENERATION

Our software architecture determines the components of a system and their translation process from specification languages to programming languages.

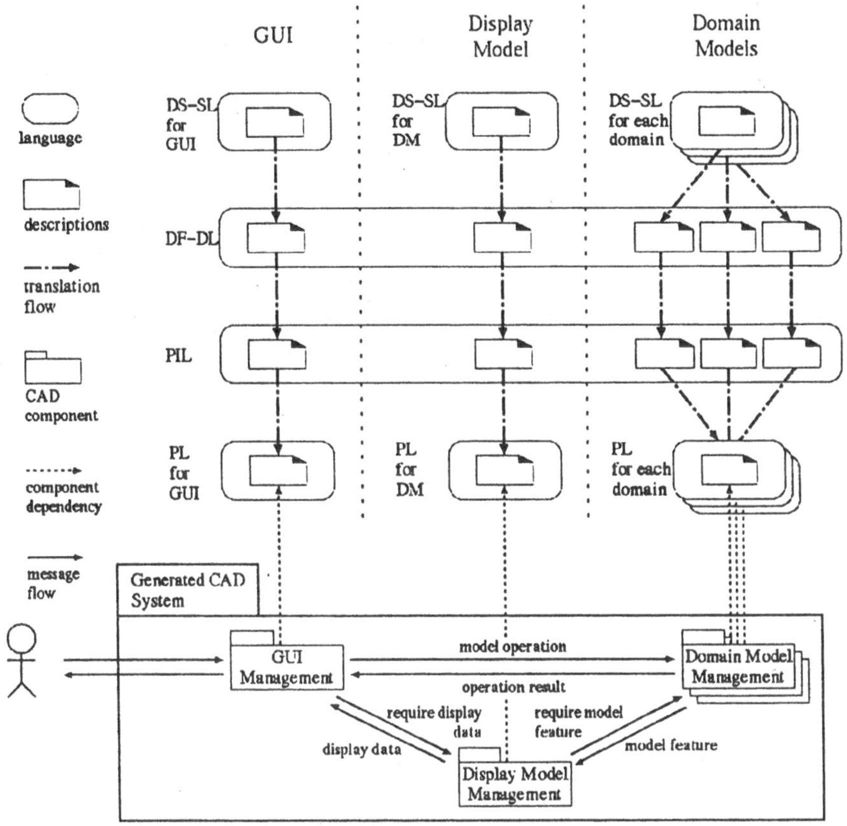

Figure 3.7 Software architecture of CAD system generation

3.5.1 Components of a CAD System

The software architecture is shown in Figure 3.7. A CAD system can be divided into three primary components based on their respective functions, each of which is expressed as a UML package (Si Alhir, 1998) at the bottom of the figure.

(1) Domain model management
This component manages the product data of the domain by using the various constraints to be satisfied among the data.

(2) GUI management
This component is responsible for supporting an interface for users. It receives operation commands from users and provides methods that allow the users to generate, delete, and modify object data.

(3) Display model management
This component supports a method of displaying object data as defined and modified by the domain model management.

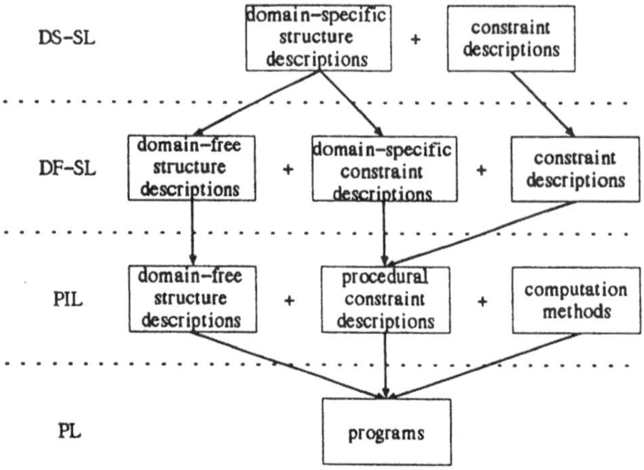

Figure 3.8 Translation process of specifications

3.5.2 Translation Process

Specifications of CAD tools are classified into the following:

(1) structure descriptions,
(2) constraint descriptions.

Therefore, we translate the specification descriptions in two phases, translation of structure descriptions and that of constraint descriptions.

As discussed in 3.3, there are many kinds of data models for CAD tools. However, we have selected the ER model as an intermediate model to generate CAD tools, and specified a domain-free specification language (DF-SL) based on the ER model. As we suppose the specifications of CAD tools are written in domain-specific specification languages (DS-SL), structure descriptions in DS-SL are translated into the structure descriptions in DF-SL in the first phase of the translation. In this phase, the constraint descriptions are not substantially translated, but the attributes of the domain specific model are newly added to the constraints.

When translating constraint descriptions to procedures, we must not only translate the constraints but also add computation methods as described in 3.4. The target language of this translation is not a specific programming language (PL), but a procedural intermediate language (PIL), which is based on an object-oriented programming language. Using the PIL simplifies our work to cover multiple programming languages.

Figure 3.8 summarizes of the above-mentioned translation process. As the translation is applied to each component of the CAD tool, the whole translation process becomes as shown in Figure 3.7.

3.5.3 Control Method of CAD Systems

All the components of CAD systems are controlled by the methods appropriate for achieving their imposed functions. These control methods are classified into the following three level groups in our software architecture.

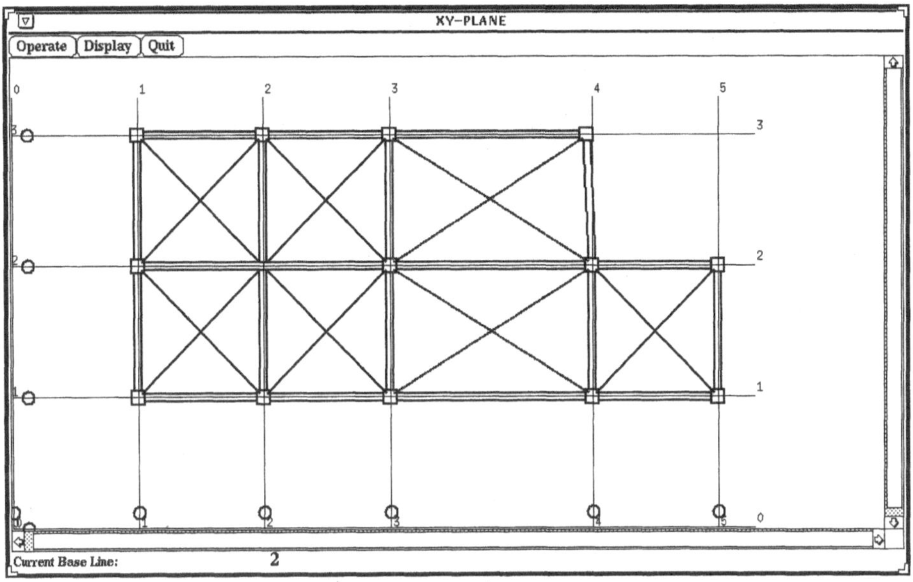

Figure 3.9 GUI example of IBDS

(1) Command menu level
Each design function starts when a designer selects a command menu button in a display window to raise an event. Therefore, the event-driven method is applied to the GUI management component. These highest-level methods refer to the lower level methods.

(2) Model-driving level
The domain model management component should maintain consistency among the product data, which have constraints. Therefore, it should be able to revise the data when a contradiction is discovered. The TMS (Truth Maintenance System) (Doyle , 1979) provides a technique for choosing among actions when a contradiction is discovered in a data set. Employing the TMS can reduce calculation when a change is performed, because it allows recalculation only in the affected range of that change. The display model management component is responsible for expressing how data will be displayed. It must receive data changes immediately from the domain model management. Therefore, the component should also adopt the TMS.

(3) Basic level
Basic level methods provide functions, such as creation and deletion of objects, and the setting of attribute values, which constitutes direct manipulation of the objects in the CAD systems. The methods in this level are automatically embedded to each component.

3.6 PROTOTYPE IMPLEMENTATION

In the sample domain of architecture, a domain-specific specification language BDL has been defined. Based on the BDL, we have already developed a system generator for IBDS (Integrated Building Design System), shown in Figure 3.9 as an implementation prototype. This section outlines BDL as the DS-SL, along with other languages defined in the prototype, and describes the system generator and its experimental applications.

```
PART column
 PLUG
  lowercColumn INTO column::upperColumn;
 END
 SOCKET
  uppercColumn TAKE column::lowerColumn;
  westBeam TAKE eastWestBeam::eastDependentPart;
  ....
 END
 ATTRIBUTE
  weight REAL(kg) :- DEFAULT(400);
  load REAL(kg) :-weight +upperColumn->load + (westBeam->load + eastBeam->load
                    + northBeam->load + southBeam->load) / 2;
  ...
 END
END PART
PART eastWestBeam
 ...
END PART
```

Figure 3.10 Example of BDL descriptions

3.6.1 DS-SL

Figure 3.10 shows an example of the specifications of the "column" part, using the domain-specific BDL. The language is based on the ER model and imposes a restriction on the association as one-way binary links. The BDL lists the names of parts, and the names of their attributes. Concerning attributes, the BDL describes data types, measures, and default values. Moreover, the BDL lists existence dependency (ED) and attribute dependency (AD) constraints.

The ED constraint is a one-way binary association that connects two parts. The constraint is described using "PLUG" and "SOCKET" statements in the figure. The load stream flows from a "PLUG" side part to a "SOCKET" side part. In the figure, a column gives its load to the lower column. In contrast, it must take the load from the upper column and the connected beams.

The AD constraint determines the computation of attribute values. In the Figure, a "load" value of a column is obtained from the load of the upper column and the connected beams using "+" and "/" calculation operators. In the statement, "upperColumn->load" specifies the "load" value of the "upperColumn," where "upperColumn" is a SOCKET name.

3.6.2 DF-SL

In Figure 3.11, the DF-SL descriptions are shown in the predicate form of the intermediate language. The DF-SL uses the same meanings as these in the ER model. For example, "dfsl_class" with "e" in the head of the bracket expresses an entity, and "r" describes a relationship. The name of the association is generated in the system generator because the name is not given in the BDL. "dfsl_attribute" expresses an attribute, whereas "dfsl_computeAt" and "dfsl_constraintParam" describe a calculation expression of AD and its parameters, respectively. A pair of "dfsl_collection" specifies two entities related by a relationship, and includes the names of their roles in the relationship. The role names are described as "PLUG" and "SOCKET" names in the BDL.

```
dfsl_class(e,'column',21,us).
dfsl_attribute('column',21,'weight',1,us,400,real,'kg').
dfsl_attribute('column',21,'load',2,d,['nil','this'],'real','kg').
dfsl_computeAt('column',280,'','this','load','V1+V2+(V3+V4+V5+V6)/2').
dfsl_constraintParam(284,'V1','','this','weight',nil).
...
dfsl_class(r,'lowerColumn_column_upperColumn_column',62,us).
dfl_collection('lowerColumn_column_upperColumn_column','upperColumn','column',plug).
...
```

Figure 3.11 Example of DF-SL

3.6.3 PIL

The PIL descriptions are shown in Figure 3.12. The language is procedural, and is based on an object-oriented language, such as C++, as an objective program. In the figure, "pil_class" and "pil_att" describe a class and an attribute, respectively, as data structure. "pil_classUsage" and "pil_attUsage" add supplementary information to the class and the attribute, respectively.

Our software architecture has three levels of control methods, as mentioned in 3.5.3. At the highest command menu level, various methods for design function entries are attached to the command menu objects.

The model-driving level of the prototype implements the TMS algorithm for maintaining the consistency among data to realize the functions mentioned in 3.5.3. Our representative implementation of methods for the TMS is shown in Figure 3.12 as "updTmsAtt" and "getTmsAtt." The former maintains the consistency of attribute values by updating them, in response to a changed attribute value, with the special denotation "nil" specifying the attributes that will be unavailable. The latter initiates a recalculation to obtain a new valid value for such an attribute.

The basic level methods, which are not illustrated in the figure[2] directly,

```
pil_class('column').
pil_classUsage('column',b,0,m).
pil_mtd(newObj,'11','column','',v0,['OutObjPtr'],
        seq([exc(newObj,'11','column','',['OutObjPtr'])])).
...
pil_att('column','weight',real,400).
pil_attUsage('column','weight',us,kg).
pil_mtd(getTmsAtt,'11','column','weight',v0,['ObjId','V'],
        set([acs(getDrtAtt,'11','column','weight',['EntityId','V1'])],
            [selExpr('V1\\==nil')], [selExpr('V=V1')],
            [acs(attDepn,'11','column','weight',['EntityId','V2']),selExpr('V=V2'),
             acs(setDrtAtt,'11','column','weight',['EntityId','V'])])).
pil_mtd(updTmsAtt,'11','column','weight',v0,['InObjId','V1'],
        ...).
...
```

Figure 3.12 Example of PIL

manipulate objects. "newObj" and "newLink" create an object and link, and "delObj" and "delLink" delete an object and link, respectively. "getObj" obtains an object related to the original object, and "getLink" gives the link between them. Concerning an attribute, "getDrtAtt" (get directly attribute value) and "setDrtAtt" (set directly attribute value) are implemented. "attDepn" (attribute dependency) calculates AD.

The above-mentioned PIL descriptions are finally translated into the INSIDE Prolog descriptions shown in Figure 3.13.

```
class('column',super(io_parts),
  ivar(className(reader('className'),write('set_className'),initform('column')),
       'objectID'(reader('objectID'),writer('set_objetcID'),initform(nil)),
       'weight'(reader('weight'),writer('set_weight'),initform(400)),
       'load'(reader('load'),write('set_load'),initform(nil)))).
pl_computeAttribute('column',ObjectID,'load',V,No) :-
  pl_getAttribute(ObjectID,'1','weight',_,V1,_,_),
  ...
  V is V1+V2+(V3+V4+V5+V6)/2.
...
```

Figure 3.13 Example of PL

3.6.4 Experimental CAD System Application

Three applications have been developed.

(1) Fire alarm system design

We have developed an automatic designing tool for a fire alarm system (Nakashima, *et al.*, 2000). BDL is used to describe the domain model. From these descriptions, three respective functions of that system are generated. The system automatically creates rooms from given walls, and sets the fire alarm sensor at an appropriate point inside each room. Moreover, the sensors are automatically wired. The algorithm of automatic design is described using DSP (Umeda, *et al.*,1996).

(2) Architectural CALS

In an experiment conducted by the Japanese chapter of the IAI in 1998, we have successfully implemented a data conversion tool between the BDL and the IFC (Nakashima, *et al.*, 2000). The relational model belongs to the generation of data models previous to that of the object model, and lacks the concept of association present in the object model. In the experiment, this problem was one of the primary challenges in our implementation of a tool for converting data from the IFC to the BDL.

(3) Elevator Entrance Design

An entrance of an elevator is designed according to the thickness of the wall and the size of the opening. However, the CAD tool for designing a building structure is different from the one for the entrance design. We have experimented with the data conversion between these (Hirota and Hashimoto, 2001). The former CAD tool can output IFC data as described in 3.3.2, but the latter, whose model is based on an assembly structure as described in 3.3.1, cannot manipulate the IFC data. Our CAD system successfully converted the IFC data to the data specific to the latter CAD tool through DF-SL.

3.7 CONCLUSION

This paper has proposed a software architecture of CAD system generation that allows CAD system development by the domain experts themselves using a domain-specific language. This software architecture also enables us to build an integrated CAD system that contains multiple domains in a field of industry such as architecture. The resulting integrated CAD systems can introduce an object-oriented database and avoid problems relating to data exchange because they use the homogeneous ER model, which can be restricted to specific domains.

By applying this software architecture, the implementation of a prototype on a UNIX platform was performed. The prototype implementation proved that the software architecture is feasible. However, the problem of execution efficiency must be solved by applying the C++ program code generator, which we are now developing. In future study, we intend to refine the algorithm of procedure generation from the non-procedural DF-SL into PIL.

The worldwide improvement of CALS (Commerce At Light Speed) has been forcing CAD systems to support the entire life cycle of product data. On the other hand, our definition of a "domain" restricts a field of engineering to a domain with a single abstract structure. Therefore, a CAD system must support multiple domains. For example, an architectural CAD system would support multiple domains such as structure design, equipment installation design, cost estimation, and so on. The specifications of these multiple domains can be integrated by the means of the domain-free specification language.

BIBLIOGRAPHY

Abeln, O., 1997, The "CAD-Reference-Model": An Approach to Future CAD-Systems, In *CAD System Development*, edited by Roller, D. and Brunet, P., pp.3–19.

Aman, S., Miura, T., Hashimoto, M., Tegoshi, Y. and Hirota, T., 2000, An Analysis of Building Construction Process –Framework Construction–, In *FOSE 2000* (Kindaikagakusha), pp.21–27 (in Japanese).

Chen, P. P., 1976, The Entity-Relationship Model: Toward a Unified View of Data, In *ACM Trans. Database Systems*, Vol.1, No.1, pp.9–36.

Codd, E. F., 1970, A Relational Model of Data for Large Shared Data Banks, In *Comm. ACM*, Vol.13, No.6, pp.377–387.

De Martino, T., Falcidieno, B. and Haßinger, S., 1998, Design and Engineering Process Integration through a Multiple View Intermediate Modeler in a Distributed Object-Oriented System Environment, In *Computer-Aided Design*, Vol. 30, No. 6, pp. 437–452.

Doyle, J., 1979, A Truth Maintenance System, In *Artificial Intelligence*, Vol.10, pp.231–272.

Fowler, J., 1995, STEP for Data Management, Exchange and Sharing (Technology Appraisals).

Hashimoto, M. and Okamoto K., 1990, A Set and Mapping-Based Detection and Solution Method for Structure Clash between Program Input and Output Data, In *COMPSAC '90*, pp.629–638.

Hashimoto, M., Matsuoka, T., Miura, T., Gotanji, K., Hirota, T., and Katamine, K., 2000, A Discussion on Abstract Structure of Design Knowledge for Space Arrangement in Architectural Design, In *Proc. 4th JCKBSE* (IOS Press), pp.39–42.

Hirota, T., Hashimoto, M, and Nagasawa, I., 1995, A Discussion on Conceptual Model Description Language Specific for an Application Domain, In *IPSJ Journal* vol.36, no.5, pp.1151–1162 (in Japanese).

Hirota, T., 1998, Intelligent CAD to Support Structural Design of Building, in *Domain Oriented Systems Development: Principles and Approaches*, edited by K. Itoh, et al. (Gordon and Breach), pp.139–154.

Hirota, T, Katamine, K. and Hashimoto, M., 1998, Automatic Editing Operation of Building CAD, In *IDPT 1998*, Vol.4, pp.126–133.

Hirota, T. and Hashimoto, M., 2001, Data Model Transformation in CAD System for Multi-View Modeling, In *Wuhan University Journal of Natural Sciences*, Vol.6, No.1–2, pp.410–415.

International Alliance for Interoperability, 1997, Industry Foundation Classes – Release 1.5 Model Reference Documentation Final Version.

Itoh, K., Hirota, T., Kumagai, S. and Yoshida, H., 1998, Domain Oriented Systems Development: Principles and Approaches, in *the series of Advanced Information Processing Technology*, Vol.1 (Gordon and Breach Science Publishers).

Jackson, M. A., 1975, Principles of Program Design (Academic Press).

Katamine, K., Umeda, M., Nagasawa, I., and Hashimoto, M., 2001, Integrated Development Environment for Knowledge-Based Systems and Its Applications, In *SNPD'01*, pp.739–745.

Miura, T., Hashimoto, M., Tegoshi, Y., Hirota, T., and Katamine, K., 2000, A Product-Centered Process Modeling of Building Construction, In *FOSE 2000*, (Kindaikagakusha), pp.13–20 (in Japanese).

Miyazaki, Y., Katamine, K., Hirota, T. and Hashimoto, M., 1999, A GUI Generator for Domain Specific CASE Tools, In *IDPT 1999*, pp.49–56.

Nakashima, T., et al., 2000, An Experience Report on an Implementation of Architectural CALS, In *IDPT-2000*, pp.129–136.

PTC, http://www.ptc.com/products/proe/index.htm

Rumbaugh, J., Blaha, M., Premerlani, W. Eddy, F. and Lorensen, W., 1991, Object-Oriented Modeling and Design (Prentice-Hall).

Shin-yashiki, Y., Hashimoto, M., Hirota, T., Katamine, K. and Tegoshi, Y., 1998, A Case Study on Architectural Construction Process Modeling – Process Aggregation and Constraints –, In *Proc. 3rd JCKBSE* (IOS Press), pp.299–302.

Si Alhir, S., 1998, UML: A Desktop Quick Reference (O'Reilly).

Umeda, M., Nagasawa, I. and Higuchi, T., 1996, The Elements of Programming Style in Design Calculations, In *9th IEA AIE*, pp.77–86.

CHAPTER 4

Realization of Fully Independent Component Architecture

Hironori Washizaki

Yoshiaki Fukazawa

Waseda University

ABSTRACT

The Graphical User Interface (GUI) application components have been widely used in specific application domains and class-level reuse. The GUI application component is a compound component, which composed of fine-grained components (GUI widgets, etc.) and specific logic. In conventional development techniques for GUI application components, the separation between the application logic part and the GUI part is incomplete inside current GUI application components. Therefore, the reusability and the extensibility of the GUI application components fabricated by conventional techniques have not been so high. In this chapter, we propose a new component architecture, "BeaM", which realizes complete separation between the application logic part and the GUI part inside the GUI application components. A development environment corresponding to the BeaM architecture has been implemented in Java language. In this chapter, we outline the concept of BeaM and the usefulness of our development environment from the domain technology.

4.1 INTRODUCTION

From the viewpoint of the domain engineering, there are two types in domains: horizontal domain and vertical domain. Horizontal domains encompass only one system part, for example, GUI, database systems, and so on (Czarnecki and Eisenecker, 2000). System parts in horizontal domains can be used at several kinds of application systems in common. On the other hand, vertical domains (also called "application domains") cover complete systems, for example, medical systems, CAD systems, and so on (Czarnecki and Eisenecker, 2000). Complete systems in vertical domains can be used only at one specific usage respectively. When developing application systems, it is important to analyze systems from both viewpoints of the horizontal domain and the vertical domain.

The GUI domain is a very important one among horizontal domains. Since the operation of interactive application systems with GUI is becoming complex, it is necessary to make elements, which compose GUI, reusable. Thus, many GUI frameworks with GUI widgets have been proposed, and have been widely used for designing and developing GUI.

Due to the success of such GUI frameworks, software component technology for the client-side development has emerged. The software component technology is based on building complex software systems from smaller components. Not only in the GUI domain, the software component technology has been a focus of attention in many horizontal domains because it is capable of reducing development costs. In particular, the use of GUI widgets and generic components with minimum logic is widespread due to the success of Rapid Application Development (RAD) tools. Many RAD tools and operating systems provide libraries of such fine-grained components (Washizaki and Fukazawa, 2002). These fine-grained components are reused at the instance level. Since it is natural to model and implement components in an object-oriented paradigm/language (Hopkins, 2000), we restrict this study to the use of the OO language for the implementation of components.

In component-based software development, new compound components are often developed by combining the fine-grained components provided by the component libraries (belong to horizontal domains) and adding specific application logic (belong to vertical domains). Since these compound components are dependent on the target application domain, we call these compound components "application components". The application components are mainly reused at the class level.

The fine-grained components are classified into visible components (widgets) and nonvisible components. In this chapter, application components that contain one or more widgets are called GUI application components. The GUI application components consist of an application logic part and a GUI part. A GUI application component is developed for use as a subsystem or a service package.

In the process of developing GUI applications, a part of the whole application is often grouped and isolated in the form of a GUI application component. Such a GUI application component plays the role of a subsystem of the application, and is often called the business component (D'Souza and Wills, 1998). Reusability is not essential for business components. However, flexible extensibility of the business components is important for their maintenance. In particular, since the design of the GUI structure is frequently changed, the widgets in the GUI part and the references to the widgets in the application logic part inside the business components should be able to be modified easily.

On the other hand, reusability is essential for the GUI application components developed as service packages. Since not all usage contexts of components can be predicted when these components are developed, it is not guaranteed that the GUI application components will be reusable without any modification. Therefore, the GUI part and the application logic part should be able to be modified or updated independently.

The reusability and the extensibility of the GUI application components can be improved by decreasing the dependence between the GUI part and the application logic part. For this purpose, the Model-View separation design pattern (Larman, 1997) and the Model-View-Controller (MVC) architecture (Bushmann *et al.*, 1999) to which the pattern is applied are widely used. In the GUI application components with the conventional Model-View separation pattern, the application logic part corresponds to the model and the GUI part corresponds to the view in the MVC architecture. However, since the references to the widgets are latent inside the application logic part, the separation between the application logic part and the GUI part is incomplete inside the conventional GUI application components. Therefore, reusing GUI application components in a certain application domain has not succeeded.

In this chapter, we propose the software architecture, BeaM (Bean for Model-View separation). BeaM can realize complete separation between the GUI part and the application logic part inside the GUI application components by separating the references to the widgets from the application logic part. As a result of applying BeaM to the GUI application components, the application logic part and the GUI part can communicate with each other without acknowledging the details of the structures of another part.

4.2 GUI APPLICATION COMPONENTS

In the development of the GUI application components, the architecture of Model-View separation is generally applied to the internal structure of the GUI application components.

4.2.1 Incomplete Model-View separation

Some RAD tools and OSs provide application frameworks by which "incomplete" Model-View separation can be realized. AWT/JFC (Pantham, 1999) in Java2SE, Document-View (Jones, 1999) in MFC and MacApp for Macintosh are typical examples. The GUI application components supplied by these application frameworks consist of the GUI part (called "GP" hereafter), which corresponds to the view of the Model-View separation pattern, and the application logic part (called "ALP" hereafter), which corresponds to the model. Figure 4.1 shows the principal relationships between the two parts, based on Unified Modeling Language (UML) (Object Management Group, 1999).

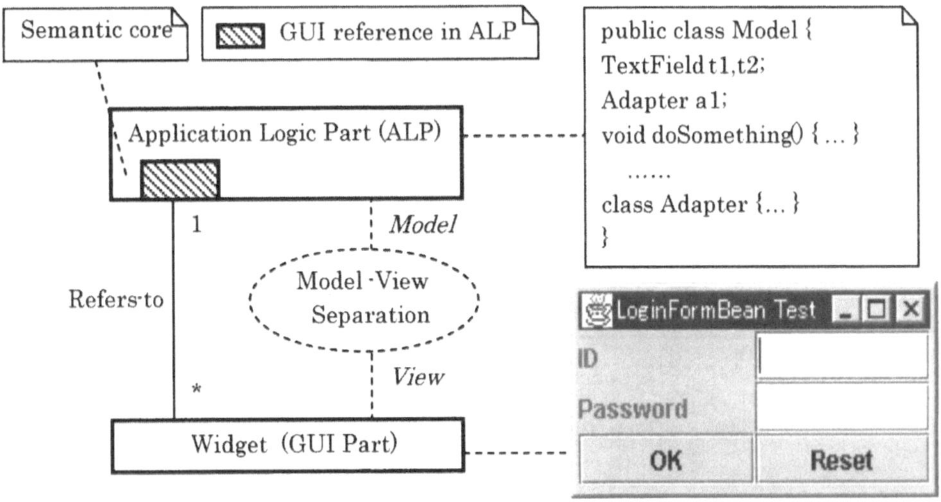

Figure 4.1 Conventional GUI application component

GUI part (GP): GP has some widgets which are provided by given component libraries attached to the RAD tools and the OSs. GP should not contain any application logic and it is possible to adjust its representation.
Application logic part (ALP): All elements of the GUI application components except for GP can be referred to as ALP. Many application frameworks are based on the

event-driven architecture where their activity is controlled by the GUI events. Therefore, ALP consists of two parts: the semantic core and the GUI reference. The semantic core contains both the operation procedures and data concerning the target application. The GUI reference contains the references to the widgets and is responsible for accepting and processing the user events from GP.

In many application frameworks, the communication between ALP and GP has been realized as follows.

(1) From GP to ALP
In the GUI application components, ALP must accept an event notification from the widgets and process it properly. In order to reduce the dependence in terms of the direction, one of the following implementations is adopted.

1. Based on the Observer design pattern (Gamma *et al.*, 1995), the implicit notification from GP to ALP and the reactive process of ALP are realized via the event adapters.

2. Based on the Template Method design pattern (Gamma *et al.*, 1995), hook methods in ALP are prepared to call back.

As an example, the mechanism of communication from GP to ALP inside the GUI application component based on AWT/JFC is shown in Figure 4.2.

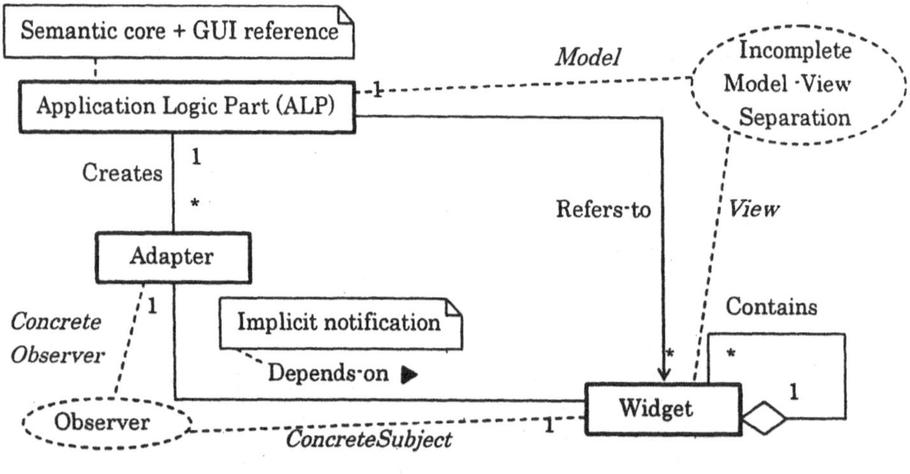

Figure 4.2 Class diagram of conventional GUI application component

In Figure 4.2, the independence of GP and ALP for the direction GP → ALP has been maintained by the implicit event notification based on the Observer pattern ([Adapter] Depends-on [View] in Figure 4.2).

(2) From ALP to GP
The data changes corresponding to the behavior of application logic in the GUI application components must be relayed to its GUI representation. In order to realize this process, ALP refers to and manipulates every widget directly in the conventional Model-View separation architectures ([Model] Refers-to [View] in Figure 4.2).

These direct references cause strong dependence between ALP and GP. This dependence is the reason why the Model-View separation provided by the conventional application frameworks is “incomplete”. This dependence causes the following serious problems.

1. Decrease of maintainability and extensibility
When a part of the GUI is changed corresponding to the specification of the GUI application component, it is necessary to change not only adapters but also the GUI reference in ALP. Since the GUI reference is latent inside ALP, it is difficult to detect the part that should be changed in ALP. This problem leads to the decrease of maintainability and extensibility.

2. Difficulty of concurrent development
When the GUI designer and the application designer are different, GP and ALP cannot be concurrently designed or developed in either the design or development phase of the GUI application components.

4.2.2 Toward complete Model-View separation

There has been some research on excluding the mutual dependence between ALP and GP by separating the GUI reference from ALP. Pluggable MVC (Krasner and Pope, 1988) in Smalltalk and the additive UI framework Jaguar (Lovejoy and Mervine, 1999) are examples of separating GUI references.

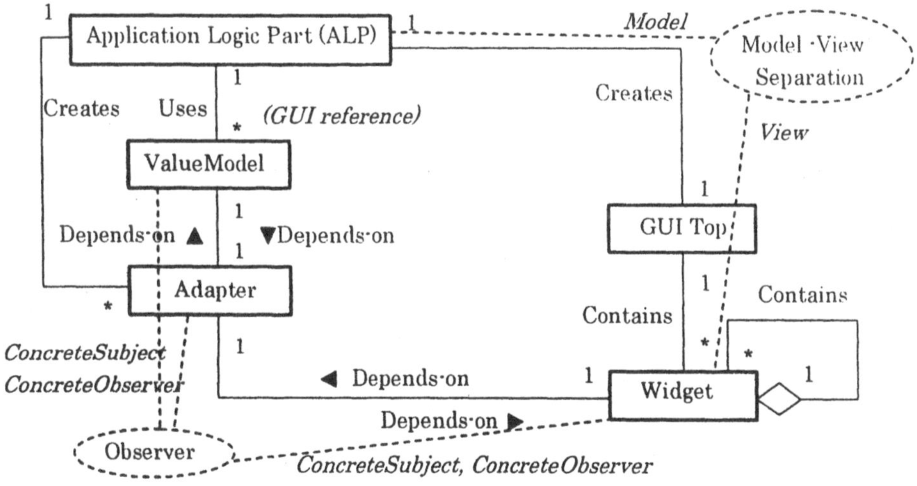

Figure 4.3 Class diagram of complete Model-View separated GUI application component

These approaches separate the model into the semantic core and the model adapters (ValueModel). The model adapters correspond to the GUI reference in the conventional Model-View separation architecture. The targets of these approaches are to realize complete Model-View separation by applying the Observer pattern to both directions of communication between the semantic core and the widgets (shown in Figure 4.3).

In these approaches, the dependence on the ALP-to-GP direction has been excluded by the one-to-one correspondence between the data items in ALP and the properties of the widgets ([ValueModel] Depends-on [Adapter] Depends-on [Widget], [Widget] Depends-on [Adapter] Depends-on [ValueModel] in Figure 4.3).

However, the one-to-one correspondence between the data items and the properties leads to an increase in the number of adapters. The correspondence also makes the message flow much more complicated in the GUI application components.

As a result, the parts that should be modified in the GUI application components lie scattered over many adapters when the specification is changed (Vlissides, 1996). This is a very weak structure for future changes in the dependence between the semantic core and the widgets. In particular, the addition of a new GUI or a change in the semantic core involves high cost.

4.3 BEAM ARCHITECTURE

In order to realize complete separation between ALP and GP in the GUI application components, we propose the BeaM software architecture. BeaM is the additional architecture for the GUI application components based on the component systems that can treat the widgets and support the mechanism of event notification. ActiveX and JavaBeans are component systems that are suitable for treating widgets and supporting the mechanism (Szyperski, 1997).

4.3.1 Design principle of BeaM

Name notification", "central management", and "external forwarding" are the fundamental design principles of the BeaM architecture.

Name notification: Similar to the conventional Model-View separation architecture, the Observer pattern with the event adapters is applied to both directions of communication between ALP and GP. There are two ways of implementing the Observer pattern (Gamma *et al.*, 1995): the push model and the pull model. In the push model, the observable sends observers detailed information concerning change. In the pull model, the observable sends nothing but the most minimal notification, and observers ask for details explicitly thereafter. Since the pull model is superior in terms of reusability compared with the push model, we adopt the pull model as the communication method between ALP and GP. The simple "name" is used as the minimal notification information.

Central management: The dependence between ALP and GP is excluded by introducing the name notification mechanism. However, as a result, the number of adapters will be excessive. Therefore, we apply the Mediator pattern (Gamma *et al.*, 1995) in order to prevent the internal structure from becoming complicated. All event adapters are brought together as one mediator. The mediator has the name conversion table to convert the notified name into a name that can be interpreted by the observer.

External forwarding: The GUI application component that consist of some components should behave, externally, as one component in the target component system. Therefore, we apply the Facade pattern (Gamma *et al.*, 1995) in order to treat the GUI application component as one component. ALP and GP have the methods that should be able to be invoked from the outside. Then, the external behavior can be realized by implementing those methods in the mediator in the central management mechanism and forwarding the accesses to the mediator from the outside to ALP or GP.

4.3.2 Basic structure

Based on the fundamental design principles, the GUI application component to which BeaM architecture is applied (called "BeaM component" hereafter) consists of the following three parts: Model, UI and Container (Figure 4.4).

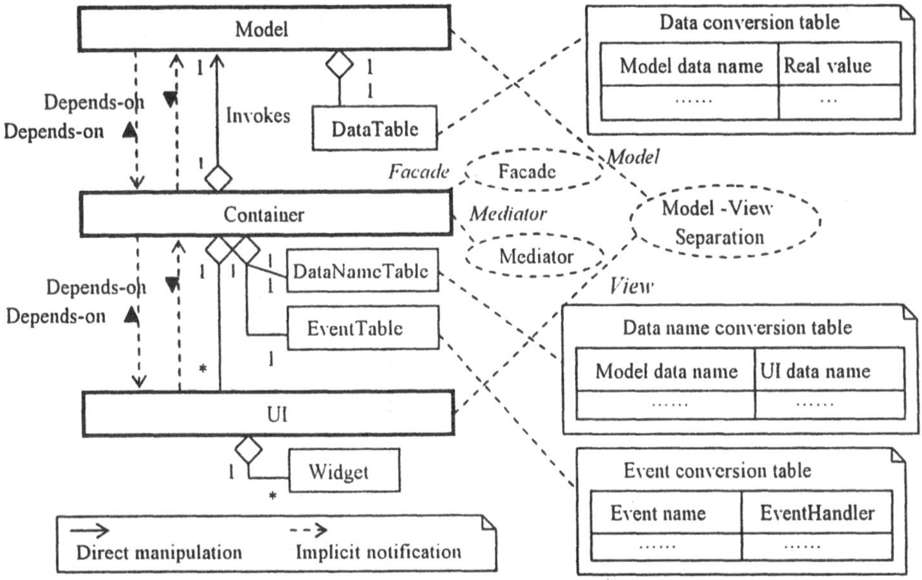

Figure 4.4 Class diagram of BeaM architecture

Model: Model contains the data conversion table and the operation procedures as the application logic. The data conversion table has sets of pairs of the real data value and the model data name. The model data name is a unique name in Model that identifies the real data value. Model communicates with Container using the model data names. Model corresponds to the semantic core in the conventional Model-View separation architecture.
UI: UI contains some widgets that are provided by given component libraries based on the target component system. UI provides the GUI representation as the result of the behavior of the entire BeaM component using those widgets. Moreover, UI accepts the input from the user via the widgets. UI communicates with Container using the event names and the ui data names. The ui data name is a unique name in UI that identifies the property name of the widgets. The event name is a unique name in UI that identifies the event that is fired by the widgets with the user actions. UI corresponds to GP in the conventional Model-View separation architecture.

Container: Container has one Model object and one or more UI objects internally. When Container has two or more UI objects, Container, as the GUI application component, provides two or more views in terms of the same model at the same time. Container has the data name conversion table and the event conversion table. Container corresponds to the mediator and the facade in the fundamental design principles. The data name conversion table has sets of pairs of the model data name and the ui data name. The event conversion table has sets of pairs of the event name and the event processing procedure. The event processing procedure consists of the operation procedure of Model and the ui data names corresponding to the inputted data. The operation procedure is implemented as the command object that refers to the method of Model by applying the Command pattern (Gamma *et al.*, 1995). The content of the data name conversion table and the event conversion table can be dynamically changed. Container corresponds to the event adapters and the model adapters in the conventional Model-View separation architecture.

4.3.3 Communication method

All communications between Model and UI are performed through Container in order to retain the independence of Model and UI for both directions. Moreover, all communications of the entire BeaM component with the outside are performed through Container.

(1) Data sharing (Model ←→ Container ←→ UI)
In the design phase, application programmers determine the data that should be shared with UI among data in Model. BeaM architecture supplies an automatic data-sharing mechanism by means of name notification based on the pull model of the Observer pattern for both directions ([Model] Depends-on [Container] Depends-on [UI], [UI] Depends-on [Container] Depends-on [Model]). The real value of data in Model and the value of the GUI representation by UI easily become equal via this data-sharing mechanism.

The flow of data sharing is described as follows. First, the shared data in Model forms a pair with the model data name when its value is initialized. The pair of the real data value and the model data name is registered in the data conversion table. Second, when the data value is changed as a result of the behavior of the application logic in Model, Model notifies Container of the model data name corresponding to the changed data. Third, Container converts the notified model data name into the ui data name using the data name conversion table. Container notifies all UIs of the ui data name. Fourth, all UIs extract the altered value of data using the ui data name from Model via Container. Finally, all UIs update the property value of the widget corresponding to the notified ui data name with the extracted new value of data.

In contrast, when the user inputs a new value to the widgets in UI, UI notifies Container of the ui data names corresponding to the widget's properties, the values of which were input by the user. Container converts the notified ui data names into the model data names and notifies Model of the model data names. Model extracts the value of data input using the model data name from UI via Container.

(2) Internal event transaction (Model ← Container ←→ UI)
All events according to the user input are processed in the BeaM component itself. When a user event occurs in UI, UI notifies Container of the event name corresponding to the event that is fired by the widgets with the user actions. Container extracts the event processing procedure using the notified event name from the event conversion table.

First, Container extracts the ui data names from the extracted event processing procedure. Second, data sharing in terms of the entered values in UI is processed (as described in Section 3.3(1)). Third, Container extracts the operation procedure from the extracted event processing procedure. Fourth, Container automatically invokes the Model method using the extracted operation procedure ([Container] Invokes [Model] in Figure 4.4). Although Container directly manipulates the Model methods, the target method has not been set statically in Container. Therefore, Container does not depend on any specific method or any specific Model.

(3) Communication with outside
Model and UI include the methods that should be able to be invoked from the outside. Therefore, Container behaves as one component by implementing those methods to Container and forwarding the access to Container from the outside to Model and UI.

4.3.4 Implementation on JavaBeans

We have implemented BeaM architecture on JavaBeans (Brookshier, 1997), which is a component system based on Java language. Figure 4.5 shows the principal relationships among the classes of the implementation of BeaM.

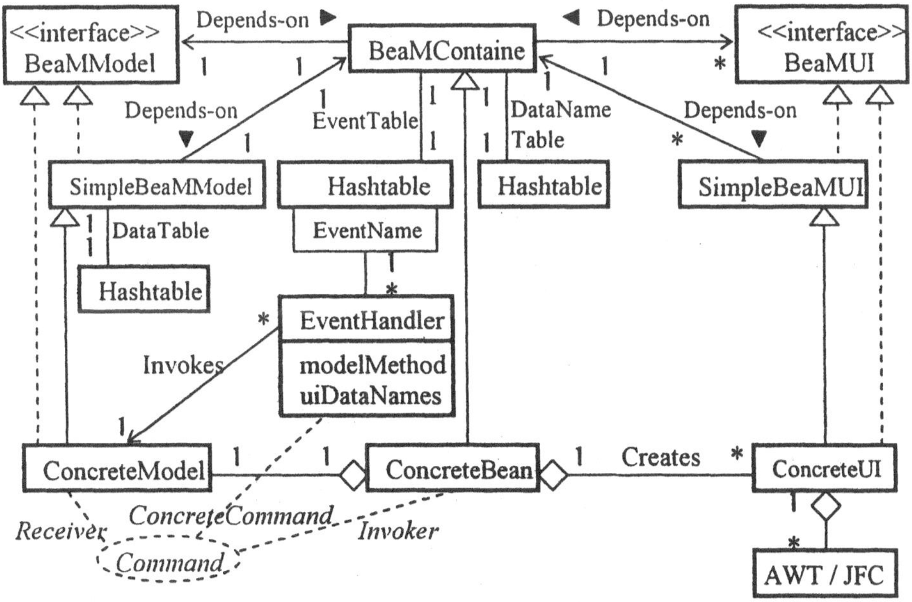

Figure 4.5 Class diagram of BeaM component

BeaMModel: The BeaMModel object corresponds to Model in the BeaM architecture model. The BeaMModel object adapts the Java's java.util.Hashtable object as the data conversion table.

BeaMUI: The BeaMUI object corresponds to UI in the BeaM architecture model. The BeaMUI object has some AWT/JFC components as widgets, and represents the GUI of the entire GUI application component.
BeaMContainer: The BeaMContainer object corresponds to Container in the BeaM architecture model. The BeaMContainer object has one BeaMModel object and one or more BeaMUI objects internally. The BeaMContainer object also has two Hashtable objects as the data name conversion table and the event conversion table. The event processing procedure in the BeaM architecture model is represented as the EventHandler object in the implementation. The EventHandler object has the BeaMModel method object as the operation procedure. The EventHandler object also has an array of String objects as the ui data names. The external behavior as a JavaBeans component is realized as an access transfer to properties of the BeaMModel object or the BeaMUI object from the outside of the component and the external event notification to the outside via the BeaMContainer object.

4.4 DEVELOPMENT ENVIRONMENT

4.4.1 Development model

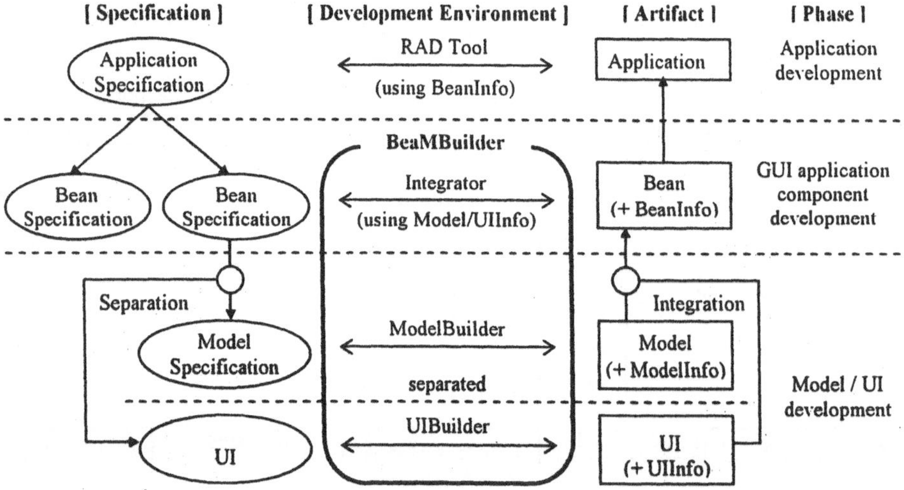

Figure 4.6 GUI application component development model with BeaMBuilder

Concurrent development of each element based on the complete separation of Model and UI is one of the purposes of applying the BeaM architecture to the GUI application components. Therefore, before the GUI application component is developed, it is necessary to separate the specification of the GUI application component to the Model part and to the UI part. This specification separation can be achieved by, for example, the

Visual Modeling Technique (VMT) (Tkach et al., 1997). In the analysis phase using VMT, use-case modeling and rapid prototyping of the view can promote the separation of the specification and the following separated developments. After the specification is separated, the Model part and the UI part can be independently implemented using the development tools. Figure 4.6 shows the flow of this development model.

4.4.2 BeaMBuilder: Integrated development tool

Since components are generally integrated visually using RAD tools which are based on the WYSIWYG philosophy (Brookshier, 1997), a tool by which the GUI application component can be developed visually along with the BeaM architecture is indispensable. Therefore, we have implemented the integrated development tool "BeaMBuilder". BeaMBuilder is a tool for developing the BeaM components in Java language. BeaMBuilder consists of three tools: ModelBuilder (for BeaMModel), UIBuilder (for BeaMUI), and Integrator (for BeaMContainer).

ModelBuilder: ModelBuilder helps programmers to develop the subclass of the class BeaMModel by coding and partial visual setting. Using ModelBuilder, programmers can visually determine the model data names corresponding to the data whose value should be shared in the final BeaM component. ModelBuilder generates the program codes of the Model class and stores the information concerning the shared data in the ModelInfo object.

UIBuilder: UIBuilder helps GUI designers/programmers to derive the subclass of the class BeaMUI by designing GUI visually. Using UIBuilder, programmers can visually determine the ui data names corresponding to the widgets' properties, the values of which should be shared in the final BeaM component. Programmers can also visually determine the event names corresponding to the events that are fired by the widgets with the user actions. UIBuilder generates the program codes of the UI class, and stores the information concerning the shared data and the internal events in the UIInfo object.

Integrator: Integrator helps programmers to develop the subclass of the class BeaMContainer by partial visual setting. Using Integrator with the ModelInfo object and some UIInfo objects, developers can visually determine the connections between the model data names and the ui data names for data sharing. The programmer can also visually determine the BeaMModel object's methods corresponding to the internal event that will be notified by the BeaMUI object. Integrator generates the program codes of the BeaMContainer class as the BeaM component, and stores the information concerning the properties and the external events in the BeanInfo object. As a result of generation, the final BeaM component can be recognized as the JavaBeans component by other RAD tools, which support JavaBeans.

4.5 EVALUATION

By developing LoginFormBean as an example, we will demonstrate the technique of developing the GUI application components based on the BeaM architecture using BeaMBuilder, and will also show the advantages of modifiability and extensibility compared with the conventional technique. We use JBuilder2.0 (Borland Software Corporation, 2002) as a comparative target. JBuilder supports programmers developing the GUI application components based on the incomplete Model-View separation.

LoginFormBean is the JavaBeans component with which the functions of AuthenticatorBean (Sun Microsystems, 2002) are simplified (Figure 4.7).

LoginFormBean is used for entering the user ID and password. LoginFormBean has "ID" and "password" as its properties, and the methods of accessing them (getId, setId, getPassword, setPassword) are made public.

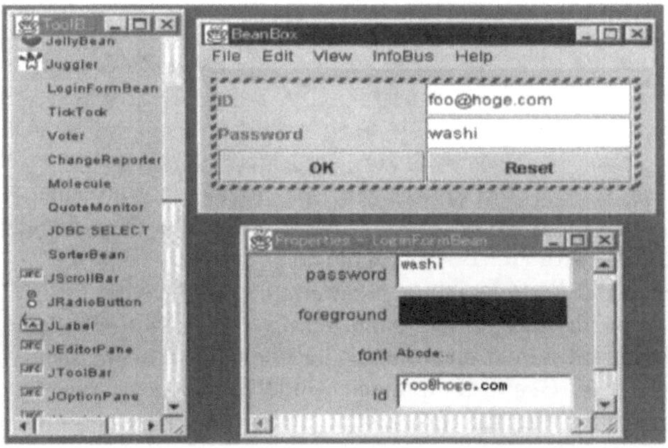

Figure 4.7 Example of display of LoginFormBean using BeansDevelopmentKit

4.5.1 Detailed customization in each part

When the requirement specification of LoginFormBean has been changed, LoginFormBean is modified to adapt to this specification change. The GUI definition of LoginFormBean is assumed to be changed: JTextField for the password is changed to JPasswordField, and JTextField for the ID is changed to JComboBox.

In the conventional technique using JBuilder, in ALP, programmers must modify all operation procedures in which JTextField component methods are invoked. The methods of JTextField components will be invoked inside the property access methods (setPassword, setId) which are included in ALP of LoginFormBean. Since the property access methods are essential to the JavaBeans components, the property access methods should not be modified when the GUI definition is changed. As a result, it becomes difficult to customize GUI.

In our technique, programmers only set the same ui data names as the ui data names that were set before the GUI definition was changed using UIBuilder. Programmers need not modify BeaMModel objects at all. Programmers can visually determine the ui data names for the properties of each widget using UIBuilder. Therefore, in our technique, it is easier to change the GUI definition of the GUI application component than in the conventional technique.

In Figure 4.8, we compare JBuilder with BeaMBuilder in terms of the number of modified code lines in these GUI definition changes. The result of using our technique indicates a value lower than that obtained with the conventional technique in terms of the number of lines of code that should be modified by hand coding. Moreover, no modification is necessary in the Model part (logical part).

As another specification change, the application logic in ALP or in the BeaMModel object is assumed to be changed when the "OK" button is pushed and either the ID or password is not entered; the value of the entered side is used for both the ID and password. Figure 4.9 shows the number of lines of code that should be modified to adapt to this specification change. From the ratio of the numbers of modified code lines in the conventional technique and in our technique by hand coding, 11:6, the necessary development cost which the programmer must pay can be considered to be reduced by 45.5% in our technique compared with the conventional technique. This result indicates that our technique is superior in terms of the modifiability compared with the conventional technique.

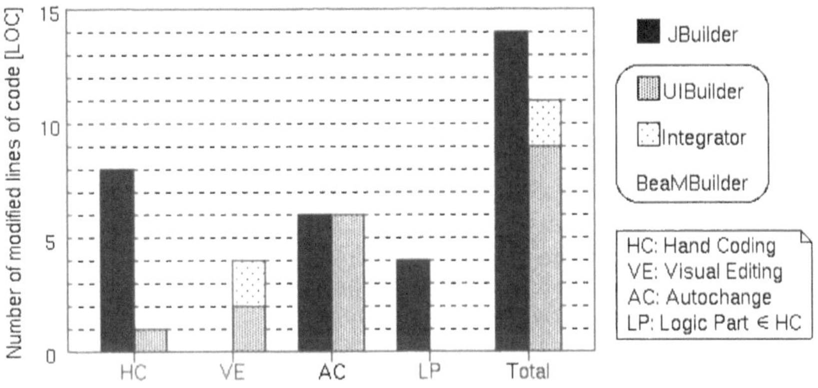

Figure 4.8 Number of modified code lines for GUI customization [LOC]

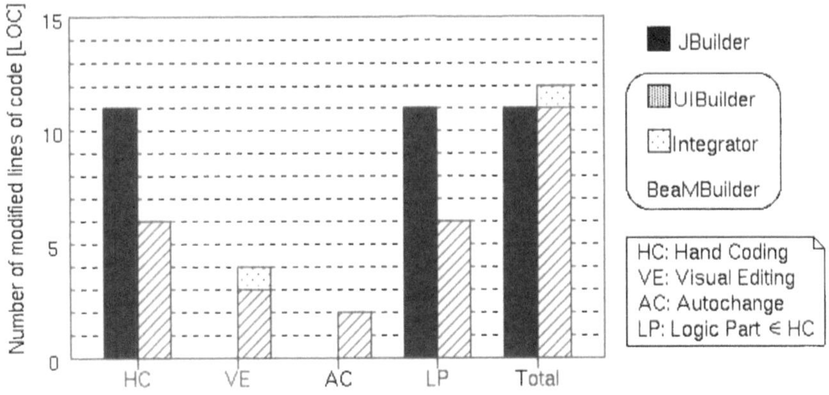

Figure 4.9 Number of modified code lines for changing model specification [LOC]

4.5.2 Data consistency

The property "password" must be updated not only when input by the user but also when set by other application logic in LoginFormBean. Therefore, in the conventional technique, programmers should carefully design and implement the interactions between the internal data in ALP and the GUI representation in GP in order to realize two that are always consistent.

In our technique, for the data whose model data name and ui data name are connected using Integrator, it is guaranteed that the value of real data in BeaMModel object and the value of the GUI representation by BeaMUI object are always the same.

4.5.3 Multiple views

In order to add the function of logging history, the LogTextArea component that provides the logging view is assumed to be added to LoginFormBean. For this change in the conventional technique, programmers must modify the access method of the property "ID" in order to relay the value of ID to the new LogTextArea inside the access method.
In our technique, the programmer should make a subclass of the BeaMUI class, which simply has the LogTextArea component, and add the connection between the model data name of the property "ID" in BeaMModel object and the ui data name that identifies the property name of the LogTextArea component inside the LogTextAreaUI, using Integrator. This does not influence the BeaMModel object at all.

Figure 4.10 shows the number of lines of code that should be modified to adapt to this view addition. Based on the result of the number of modified code lines, it is almost unnecessary for programmers to modify the BeaM component by hand in our technique. The necessary work in our technique has been performed via visual setting using UIBuilder and Integrator. Therefore, our technique is superior in terms of the extensibility compared with the conventional technique.

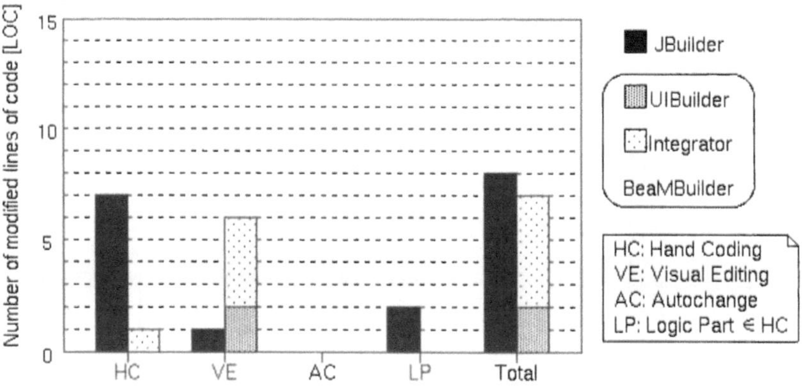

Figure 4.10 Number of modified code lines for GUI addition [LOC]

4.5.4 Runtime Performance

We verified how much the execution efficiency decreased as a sacrifice for the improvement of the modifiability and extensibility in our technique. For the LoginFormBean developed by the conventional technique and our technique, the following series of operations was automatically repeated for a specified number of times: entering values, firing the external event, and resetting values. We use the default LoginFormBean which has a single view and the extended LoginFormBean which has two views (described in Section 5.3). The execution time was measured in each repetition. Table 3.1 shows a comparison of the execution times ([ms]). In Table 3.1, the increase of execution times in our technique is 3.98% on the average compared with the conventional technique. The difference is seen to be small, so the implementation of the BeaM architecture experiences no performance problem in practical application.

Table 4.1 Execution time for LoginFormBean [ms]

	Single view		Multiple views	
Repetition [times]	Conventional	Our	Conventional	Our
100	17125	18828	18697	18787
1000	169804	176374	179588	183494
10000	1809102	1937625	2015458	2024081

(Environment: Java2 SE1.2.2, Windows NT4.0 Server, PentiumIII 500MHz, Memory 256MB)

4.6 RELATED WORKS

Presentation-Abstraction-Control (PAC) (Hussey and Carrington, 1996) architecture is based on the hierarchy of the basic agents. Each basic agent consists of three parts: Presentation, Abstraction, and Control. Through modifications of the agent and each part of the agent, PAC realizes high flexibility corresponding to any specification change. As well as the BeaM architecture, the dependence between Presentation as the view and Abstraction as the model has been excluded.

However, the implementation method of communication between Abstraction and Presentation has not been limited to the event notification style based on the Observer pattern. Therefore, programmers can develop a GUI application component with low extensibility by using PAC (Buschmann *et al.*, 1999). Moreover, similar to the complete Model-View separation architecture, PAC has a potential problem with an excessive number of adapters.

Model-View-Presenter (MVP) (Potel, 2000) architecture is based on the MVC architecture, and Presenter is an extension of Controller with new functionality for managing specific information concerning GUI representation. MVP can realize multiple views in terms of one or more models.

However, since Presenter depends on the kind of Model for the Model's GUI representation, the separation between Model and View is incomplete in terms of the direction from View to Model.

4.7 CONCLUSION

As a technique of developing GUI application components, we have proposed a new architecture, BeaM, and the corresponding development tool, BeaMBuilder. The BeaM architecture can overcome the problem of the internal dependence that existed in the conventional technique. Since the BeaM architecture is a general architecture to the component systems that can treat the widgets and support the mechanism of event notification, the architecture can be similarly applied to other component systems (ActiveX etc.)

As mentioned above, we have already developed the development environment for the BeaM architecture, and evaluated its usefulness. Although the details about the research result are described in our past paper (Washizaki *et al.*, 2001), the consideration from the viewpoint of the domain engineering was lacked. Thus, we have considered the BeaM architecture and its environment from the viewpoint of the domain engineering, and the result of the consideration is described in this chapter. In future work, we plan to continue evaluation and consideration from such viewpoints. And, we also plan to include the implementation of the BeaM architecture in other component systems.

BIBLIOGRAPHY

Borland Software Corporation., 2002, *JBuilder*, http://www.borland.com/jbuilder/

Brookshier, D., 1997, *Java Beans Developer's Reference*, (Indianapolis: New Riders Publishing).

Buschmann, F., Meunier, R., Rohnert, H., Sommerland, P. and Stal, M., 1999, *A Pattern-Oriented Software Architecture*, (Boston: Addison-Wesley).

Czarnecki, K. and Eisenecker, U., 2000, *Generative-Programming: Methods, Tools, and Applications*, (Boston: Addison-Wesley).

D'Souza, D. and Wills, A., 1998, *Objects, Components, and Frameworks with UML*, (Boston: Addison-Wesley).

Gamma, E., Helm, R., Johnson, R. and Vlissides, J., 1995, *Design Patterns: Elements of Reusable Object-Oriented Software*, (Boston: Addison-Wesley).

Hopkins, D., 2000, Component Primer, In *Communications of the ACM*, Vol. 43, No. 10, (New York: ACM), pp. 27–30.

Hussey, A. and Carrington, D., 1996, Comparing two user-interface architectures: MVC and PAC, In *Proceedings of BCS-FACS Workshop on Formal Aspects of the Human Computer Interface*, (Berlin: Springer).

Jones, R., 1999, *Introduction to MFC Programming with Visual C++*, (Upper Saddle River: Prentice Hall).

Krasner, G. and Pope, S., 1988, A Cookbook for Using the Model-View-Controller User Interface Paradigm in Smalltalk-80, In *Journal of Object-Oriented Programming*, Vol. 1, No. 3, (Chatsworth: 101communications), pp. 26–49.

Larman, C., 1997, *Applying UML and Patterns*, (Upper Saddle River: Prentice Hall).

Lovejoy, A. and Mervine, F., 1999, Jaguar: A UI Framework for Java, In *Proceedings of 14th Annual ACM SIGPLAN Conference on Object-Oriented Programming, Systems, Languages, and Applications*, (New York: ACM).

Object Management Group., 1999, *OMG Unified Modeling Language Guide Specification Version 1.4*, http://www.uml.org/

Pantham, S., 1999, *Pure JFC Swing*, (Indianapolis: Sams Publishing).

Potel, M., 2000, MVP: Model-Viewer-Presenter – The Taligent Programming Model for C++ and Java, In *developerWorks*, http://www.ibm.com/developerworks/

Sun Microsystems., 2002, *Hot JavaBeans*, http://java.sun.com/products/hotjava/bean/

Szyperski, C., 1997, *Component Software: Beyond Object-Oriented Programming*, (Boston: Addison-Wesley).

Tkach, D., Fang, W. and So, A., 1997, *Visual Modeling Technique: Object Technology*

Using Visual Programming, (Boston: Addison-Wesley).
Vlissides, J., 1996, The Trouble with Observer, In *C++ Report*, September, (New York: SIGS Publications).
Washizaki, H. and Fukazawa, Y., 2002, Dynamic Hierarchical Undo Facility in a Fine-Grained Component Environment, In *Proceedings of Fortieth International Conference on Technology of Object-Oriented Languages and Systems*, (Sydney: Australian Computer Society), pp. 191–199.
Washizaki, H., Shirogane, J. and Fukazawa, Y., 2001, The Mechanism of the Model-View Separation Architecture BeaM and Its Evaluation, In *Journal of Information Processing Society of Japan*, Vol. 42, No. 10, (Tokyo: Information Processing Society of Japan), pp. 2454–2465.

CHAPTER 5

Requirements Engineering and Domain Engineering

Atsushi Ohnishi

Ritsumeikan University

ABSTRACT

Requirements Engineering based on the Domain Analysis and Modeling (DAM), and DAM based on the Requirements Engineering are described with several examples. There exist several common processes and common languages for specifying both requirements and domain models. The author classifies DAM based on Requirements Engineering into DAM with requirements elicitation and DAM with specification techniques, and then illustrates a visual requirements language as an example of DAM based on Requirements Engineering. In Chapter 5.4, requirements definition methods based on DAM are illustrated including scenario generation method and requirements definition of jobs of program chair of an international conference.

5.1 INTRODUCTION

A couple of processes in defining software requirements are quite similar or commonly used for domain analysis and modeling. Actually, some languages and diagrams for specifying software requirements are adopted as languages and diagrams for specifying domain models.

However, the purpose of Requirements Engineering is quite different from the purpose of DAM. The goal of Requirements Engineering is both clarifying user's problems and specifying users' needs to solve the problems. Otherwise, the goals of DAM are

(1) Making domain models in a specific problem domain,
(2) Reuse of the domain models, and
(3) Effective software development with such domain models.

The differences of goals between Requirements Engineering and DAM cause to different adaptations of their methods. Here, the author describes both

(1) Supporting methods of DAM by applying Requirements Engineering techniques, and
(2) Supporting methods of Requirements Engineering by applying DAM techniques.

Supporting methods categorized into the former include acquisition of domain knowledge and interviews in domain analysis based on requirements elicitation techniques and interview techniques for requirements definition, and modeling domain knowledge with a requirements language.

On the other hand, one of the supporting methods categorized into the latter is reusing products in Requirements Engineering based on reusing of domain models. This method enables an effective software development with reusing Software Requirements Specifications in Requirements Engineering.

In Chapter 5.2, several technical terms and concepts of the Requirements Engineering will be defined, because these terms may be differently used for DAM researchers. In Chapter 5.3, DAM based on Requirements Engineering will be described. Especially a flow requirements language is introduced. In Chapter 5.4, Requirements Engineering based on DAM including a scenario analysis technique and problem specific requirements definition will be illustrated. In Chapter 5.5, the author will give concluding remarks.

5.2 REQUIREMENTS ENGINEERING

In software development, software requirements including functional requirements and non-functional requirements are firstly defined. Since a software requirements specification (SRS) is the first document in the software development and in the subsequent processes, such as software design, programming, and test, the requirements specification is regarded as a standard and basic document in the development. If users' needs are not correctly specified in the requirements specification and if the specification includes some errors, such as the inconsistency, the incompleteness, the ambiguity, developers should re-design after the delivery of software product. Such a re-design causes the delay of development schedule and the excess of estimated development cost, and sometimes causes the fatal failure of the software development project. To clarify the following discussions, the author gives definitions of technical terms in Requirements Engineering.

Requirements definition is the first process in the software life cycle, and this process where users' needs and problems are correctly defined and specified. Generally, there exist some problems and some needs in advance, and then software will be developed to meet the needs and to solve the problems. Sometimes such problems may be so complicated or so ambiguous or so inconsistent, especially in case of quite a new system. In such cases, it may be difficult for stakeholders to correctly understand their problems. In the requirements definition, problems will be analyzed and clarified, and solutions to the problems should be hit, and functional requirements and non-functional requirements must be defined to meet the solutions. Requirements definition can be divided into four sub phases, that is,

1) Collecting problems and needs
2) Analyzing the collected problems and needs, and clarifying the real problems (needs)
3) Defining the purpose of software
4) Developing a method to realize the purpose or solutions to the problem, and specifying them as a software requirements specification.

On the basis of a software requirements specification, successive processes, such as software design, software implementation, are preceded, and finally software, which realizes solutions of users' problems, will be developed.

Requirements Engineering covers several techniques and methods for eliciting, specifying, validating, verifying, managing requirements listed below (Sommerville, 1995).

Requirements Elicitation: Since users generally have different viewpoints, there exist inconsistent requirements from several viewpoints. Sometimes users do not have clarified problems, but ambiguous problems and erroneous ideas. Users' real needs should be captured and described. For requirements elicitation, there exist several methods, such as interviewing, scenario writing, KJ method, 5W1H method, problem analysis methods including brain storming, lateral thinking.

Requirements Specification Technique: Users' real needs are often described with natural language (such as Japanese and English) tables and diagrams. The described requirements is called a **requirements description**. A requirements description usually includes errors and sometimes includes infeasible requirements. Software developers should analyze such a requirements description and then should detect and correct such errors including infeasible requirements. Through these analyses, a requirements description becomes better. **Requirements specification** is a final product of requirements definition after a couple of analyses and revisions of requirements description.

Requirements language: requirements language is a language for describing and specifying software requirements. There exist several requirements languages, such as (1) formal languages, (2) control languages which are restricted natural languages in terms of restrictions of their grammar and vocabularies, and (3) graphic languages including SADT (Structured Analysis and Design Technique, currently referred as IDEF0), DFD (Data Flow Diagram), STD (State Transition Diagram), Use case diagram and other UML diagrams (OMG, 2000), and so on.

Software Project Management and requirements specification: In software development, project manager should estimate the number of developers, the development schedule, and the cost of the development. The project manager should estimate them in the early phase of the development based on software requirements specification. There exist several Software estimation methods, such as function point method, feature point method (Jones, 1996), and CoCoMo (Boehm, 1981).

Rapid prototyping: If users do not validate the correctness of software requirements specification, final product will not satisfy users' real needs. In such a case, users may not use the final product, and developers should re-develop software. Such a re-development leads to the delay of schedule of the software development and the excess of estimated cost of the software. Rapid prototyping is a method, which enables users' validating the software requirements by evaluating a prototype in an early phase of the software development. Functional behavior and user interface of the software can be validated with this method.

Supporting the requirements definition: Based on the above methods, several software tools have been developed (Davis, 1993). Such tools support the requirements definition process.

5.3 DOMAIN ANALYSIS AND MODELING BASED ON REQUIREMENTS ENGINEERING

5.3.1 DAM with requirements elicitation techniques

In domain analysis, first a domain should be identified and then a lot of information about the domain should be acquired (Itoh, 1998). Domain analysts acquire such a kind of information from several sources like technical documents and software documents and through interviews with domain experts.

Problem analysis and requirements elicitation techniques of Requirements Engineering can be useful in analyzing technical documents and software documents and in classifying them into indispensable information and unnecessary information. Both interview technique and analysis technique of meeting results are also useful in interviewing with domain experts.

5.3.2 DAM with requirements specification techniques

Acquired information about a specific domain will be represented as a domain model. There exist kinds of modeling languages for a certain domain as shown below (Itoh, 1998).

1) Entity Relationship diagram (ER diagram)
2) Class diagram
3) Data dictionary (DD)
4) Data Flow diagram (DFD)
5) Work flow diagram
6) Use case diagram
7) Structured Analysis and Design Technique (SADT, IDEF0)
8) Decision table
9) State transition diagram (STD)
10) Sequence diagram
11) Unified Modeling Language (UML) (OMG, 2000)

These languages and diagrams are mainly used for defining requirements. There exist different languages and diagrams for software design and software performance analysis, such as

1) Jackson System Development (JSD) chart
2) Hierarchical Input-Process-Output (HIPO) diagram
3) Program Evaluation and Review Technique (PERT) chart
4) Queuing network

These languages are adopted as domain modeling languages[5)].

Since domain models include not only products in the requirements definition phase, but also products in other phases of the software development, all the domain models cannot be specified with requirements languages, but such languages are actually useful to specify several domain models corresponding to the requirements definition.

5.3.3 Visual Requirements language: VRDL

The author introduces a visual requirements language named VRDL (Visual Requirements Description Language) for specifying flow requirements as an example of requirements language (Ohnishi, 1999). The features of VRDL are:

1) A requirements describer can define icons
2) He can specify data/control flow requirements with icons and arrows
3) A visual SRS is transformed into internal representation named CRD (Conceptual Requirements Description)
4) A visual SRS can be specified structurally
5) A visual SRS can be executed by giving icons' movement descriptions
6) A visual SRS can be transformed with icon dictionary

We describe the first 5 features of the VRDL.

5.3.3.1 Icon definition

Several visual requirements languages have been developed. Most of them are based on the DFD. These languages provide icons whose syntax and semantics are predefined. A describer cannot use other icons such as flow chart symbols instead. If he uses icons that faithfully reflect his original image, he can write down requirements more easily.

A VRDL user can define both the shape and semantics of icons. This feature enables the user to easily write his original image as a visual SRS. A describer can define the shape of icons as color pixmap data with existing tools, such as the ``pixmap" editor. He can also define an icon of image data.

Since VRDL is based on the Requirements Frame (Ohnishi, 1996), the semantics of an icon should be noun type provided by the Noun Frame. Namely, the semantics of an icon should be control, data, device, file, function, or human. The DFLOW concept and the CFLOW concept in the Case Frame (Ohnishi, 1996) can be drawn with VRDL, since both data flow and control flow are easily and naturally visualized. We assume that requirements about the other concepts including data structures, functional structures, and file manipulations are written with the text-base requirements language.

After the definition of the shape of icons, the describer defines the semantics of the icons and stored them in an icon dictionary. This dictionary contains icon types, icon names and icon file names. Each describer can have his own dictionary. A shared standard icon dictionary is also provided. With this dictionary, users need not define standard icons that are frequently used.

5.3.3.2 SRS with icons and arrows

A describer can write a visual SRS by placing icons and by drawing arrows. An arrow of normal line means the data flow concept. An arrow of dotted line means the control flow concept.

The describer places the icons on a sheet provided by the VRDL editor. First he selects an icon of human or function types listed in the palette of icons window as the source case object by pointing the selected icon with a mouse. Then he specifies the location of the icon with the mouse. Then he specifies the goal case object in a similar way. Next he selects the arrow command, one of the pop-up menus commands, by pointing the EDIT button in the upper side of the window and sets the starting point and the ending point of the arrow. Then he specifies a device type icon and a data type icon.

Fig. 5.1 shows a hardcopy of the editor. The upside DFLOW sentence means that a liquor shop sends an order form to a warehouse via fax. In this SRS, four DFLOW sentences are specified. These DFLOW sentences are easily transformed into CRD representation, since VRDL is based on the Case Frame.

In order to define complicated requirements, a visual SRS can be written structurally just like DFD. In Fig. 5.1, the icon, warehouse is detailed into another window labeled 1_1 to specify data flow requirements in the warehouse.

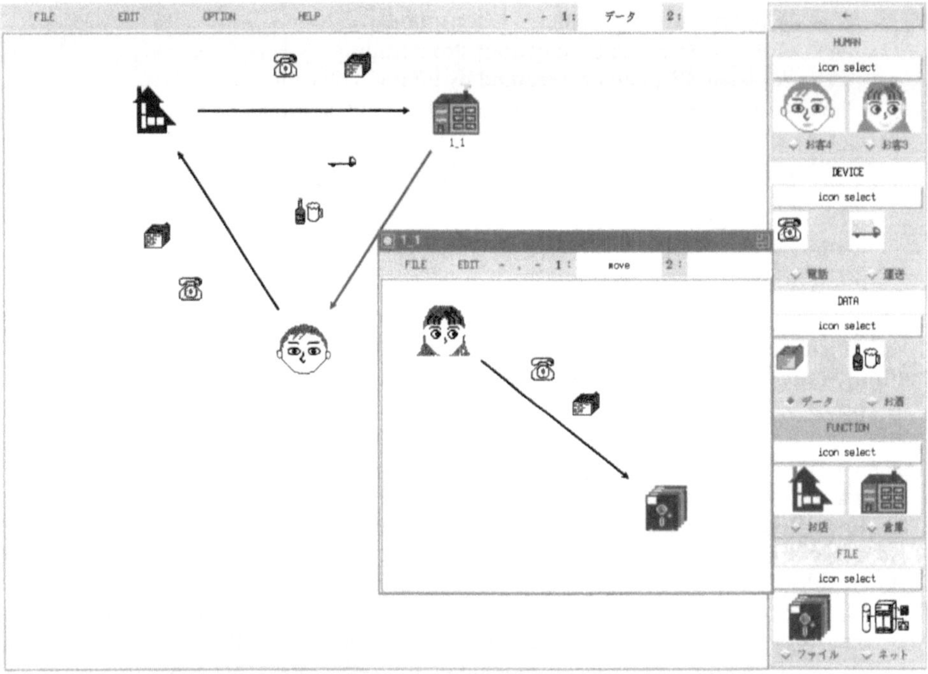

Fig. 5.1 Structured visual software requirements specification with VRDL.

5.3.3.3 Execution of visual SRS

An end user's experience with rapid prototyping is a good confirmation of the requirements. One of the rapid prototyping methods is execution of a visual SRS. The execution proposed here means icons' movement. A describer gives a scenario of icons' movements. The system behavior is simulated by interpreting this scenario description. The icons' movements are shown as animation in order that both the describer and the end user can confirm the sequence of the data/control flows.

In order to make a scenario, we provide a scenario generator. With this generator, users can specify a sequence of the data flows in a visual SRS as a scenario. By interpreting the scenario and by deriving locations of icons appeared in the scenario from the visual SRS, both data moving and the sequence of data flows can be animated. We provide two kinds of the icons' movement: one is to display and erase an icon at different locations and the other is to display a bit different icons repeatedly. For example, the movement of "data transfer" is shown as animation by the repetition of the displaying an icon at a new location and erasing the icon at the old location. The movement of "telephone's ringing" is shown as animation by switching the display of an original telephone's icon and a lifted telephone's icon.

Fig. 5.2-5.5 show snapshots of animation. The data moving and the telephone ringing are shown as animations in Fig. 5.2 and 5.3. These figures correspond to a data flow, which means that a reception clerk sends an order form to a liquor shop via fax. When the liquor shop receives an order form, another data flow defined in the liquor shop will be invoked. This situation is shown in Fig. 5.4, 5.5, and 5.6. In Fig. 5.4 and 5.5, the icon of the liquor shop is zoomed. This zooming means that data flows in the liquor shop has just begun to be animated.

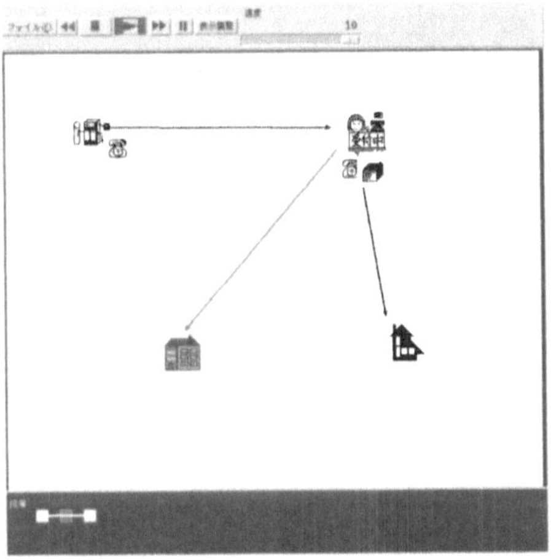

Fig. 5.2 Snapshot of the animation (1)

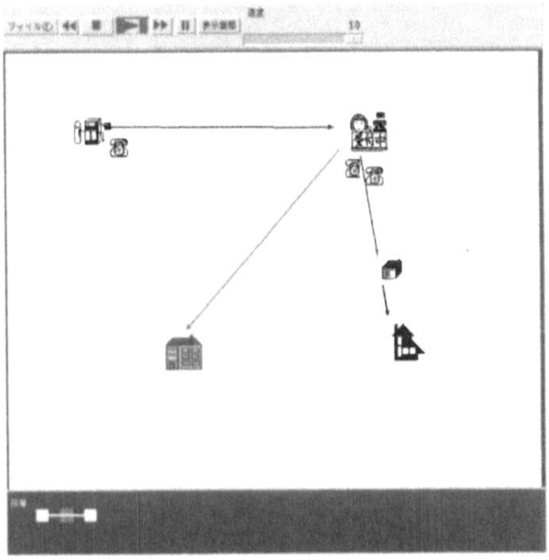

Fig. 5.3 Snapshot of the animation (2)

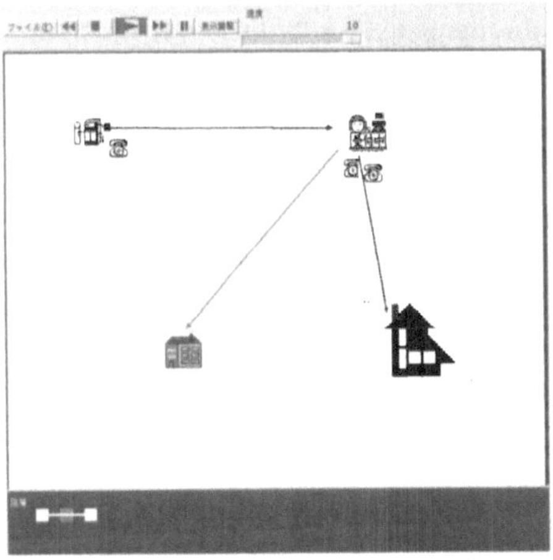

Fig. 5.4 Snapshot of the animation (3)

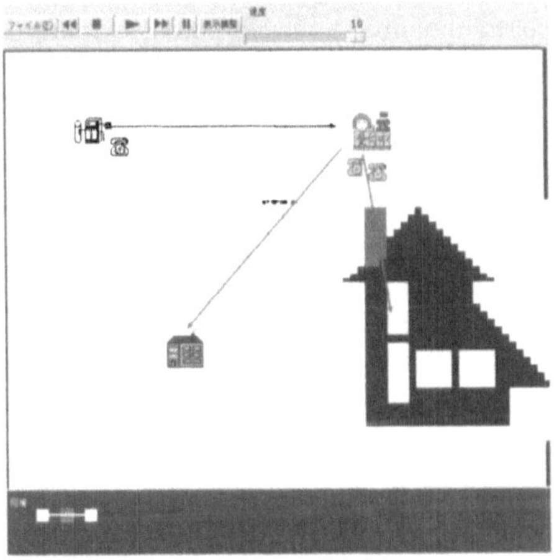

Fig. 5.5 Snapshot of the animation (4)

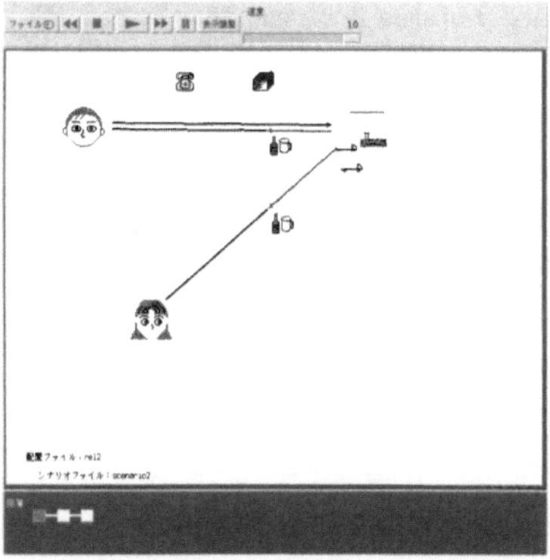

Fig. 5.6 Snapshot of the animation (5)

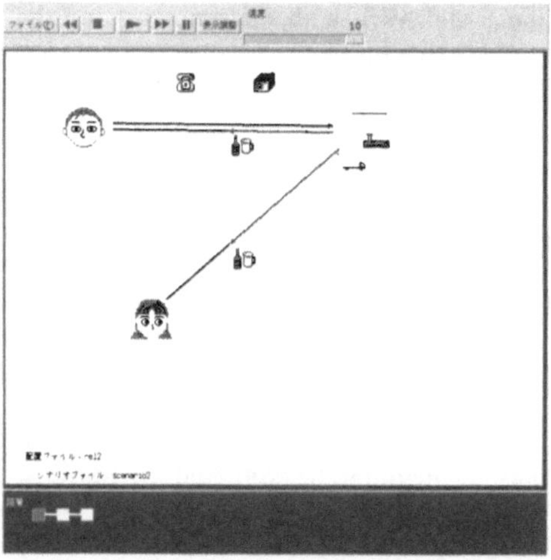

Fig. 5.7 Snapshot of the animation (6)

In Fig. 5.6 and 5.7 parallel moving of data is animated. The animation of visual specification enables user's confirmation of the directions of data moving and the sequence of data moving.

5.4 REQUIREMENTS DEFINITION BASED ON DOMAIN ANALYSIS AND MODELING

By providing several products generated in the requirements definition phase as domain models, the efficiency of the process of the requirements definition of a new system belonging to the same problem domain of the domain models can be improved. Since domain engineering covers not only requirements definition, but also design and coding phases in software development, requirements definition based on DAM will be focused on and described in this chapter. Here a common problem provided by the Working Group of Requirements Engineering (RE WG) (Ohnishi, 2001-a) is introduced. This working group is established in 1998 in Special Interest Group of Software Engineering (SIGSE), Information Processing Society (IPS), Japan. The purpose of the working group is the activation of the research area of the Requirements Engineering in Japan. There exist 21 members from 17 organizations (9 universities and 8 companies.)

The common problem will be briefly introduced in Section 5.4.1. Problem domain will be illustrated in Section 5.4.2. Scenario generation with transformation rules will be explained with lift control problem in Section 5.4.3. These topics can be classified into Requirements Definition methods based on DAM.

5.4.1 Brief introduction of the common problem

The RE working group adopted jobs of a program chairperson of an international conference as the common problem on the basis of experiences of program chairs at several conferences. This common problem consists of several jobs as shown below.

1) **Management of schedule**, such as defining the deadline of submission of papers, and the day of acceptance/rejection notification
2) **Assignment of persons**, such as keynote speakers, members of program committee, and chairperson
3) **Management of papers**, such as assignment of paper id numbers, and registration of papers
4) **Assignment of jobs**, such as assignment of a member of program committee to each of submitted papers
5) **Sending objects**, such as sending a list of papers to each of the members of program committee, sending an acceptance/rejection notification to each of the authors, and sending camera-ready papers to a publishing company
6) **Reviewing**, such as reviewing papers by reviewers
7) **Generating objects**, such as generating preface of the proceedings

5.4.2 Problem domain of the common problem

There exist similar jobs to the common problem (Tomoeda, 1999), for example jobs of program chair of a symposium and jobs of secretary of Special Interest Group's research meeting. The author has experienced jobs of program chairperson at several conferences, symposiums and research meetings, such as workshop of Foundation of Software Engineering (FOSE'98), Japan Society for Software Science and Technology (JSSST) in 1998, winter workshop of SIG Software Engineering (SIG WS), IPS, Japan in 1997 and 2001, research meetings of Requirements Engineering Working Group (RE WG) in 1998, 1999, 2000, 2001, and Object oriented symposium (oo2001), SIG Software Engineering, IPS, Japan in 2001.

The comparison between jobs of the common problem and jobs of actual meetings is shown in Table 5.1.

In Table 5.1, "A" means that "all of the jobs are quite similar to," "B" means that "a part of the jobs are similar, but the other part of the jobs are different," "C" means that "jobs are different, but some part of the jobs correspond to the jobs of the common problem," and "N" means that "there are no corresponding jobs." "Other jobs" means that there do not exist in the common problem, but these jobs are included in some conferences. In FOSE'98, the program chair should make hotel booking and assignment of rooms, although a local chair should usually do such jobs in many other conferences.

In case of FOSE'98 workshop and oo2001 symposium, most jobs of the common problem can be reused, because such conferences include jobs of making call for paper, reviewing submitted papers, making camera-ready paper for each of accepted papers, presenting at the conference.

In case of SIG WS, there exists a job of reviewing submitted papers. However, program chairs do not adopt reviewers, but review the papers by themselves. So, the jobs of assignment and reviewing are different from the jobs of the common problem. In case of RE WG, since there do not exist the jobs of submission of papers and reviewing of the submitted papers, these jobs are not found in the jobs of the chair at RE WG.

The jobs of prize essay contest, programming contest, and piano contests are quite similar to the common problem, because these jobs include holding a prize contest for something, reviewing submitted objects with a some guideline, give a prize to reviewing results as shown in Table 5.2. Major difference between such a contest and the common problem is the timing of reviewing: in a piano contest, playing techniques and expression techniques of the piano are reviewed. Playing the piano and reviewing are simultaneous. In the common problem, however, a camera-ready paper will be published by revising the submitted paper with reviewers' comments.

Table 5.1 Comparison between jobs of the common problem and jobs of actual meetings.

Job	FOSE'98	SIG WS	RE WG	oo2001
Management of schedule	A	B	N	A
Assignment of persons	A	B	N	A
Management of papers	A	A	N	A
Assignment of jobs	A	B	C	A
Sending objects	A	A	N	A
Reviewing	A	B	N	A
Generating objects	A	A	N	A
Other jobs	Hotel booking, Room assignment, Registration, Invoice, Receipt	Registration	Receipt	WWW pages, Camera-ready format

Table 5.2 Comparison between the common problem and jobs of contests.

Job	Essay contest	Piano contest	ACM programming contest (ICPC)
Management of schedule	A	A	A
Assignment of persons	A	A	A
Management of objects	A	A	A
Assignment of jobs	A	B	A
Sending objects	A	N	B
Reviewing	A	C	C
Generating	B	B	B

In the ACM International Collegiate Programming Contest (ICPC), sixty-four teams of students advancing from around 3,000 teams representing more than 1,000 universities in 70 countries on six continents competed for bragging rights and prizes at the World Finals of the annual ICPC. Student's making programs are automatically reviewed with computers by checking input and output data to the programs.

Through the above discussions, the following fact is clarified, that is, many parts of software requirements specification of the common problem and other products in the requirements definition of the common problem can be reused in developing a system for a similar problem such as FOSE'98 workshop, object-oriented symposium, and contests.

A couple of detailed jobs of the common problem, such as management of schedule and assignment of persons are frequently appeared in more general jobs. By modeling of requirements of such detailed jobs as domain models, domain models of the detailed jobs of the common problem can be reused in more general system developments.

5.4.3 Scenario and its transformation

In this section, we outline a frame-based approach for structuring the actions in a scenario (Ohnishi, 2001-b). We use ideas from the previous work of the Software Requirements Database (Ohnishi, 1996) and Jackson's problem frames (Jackson, 1995) for structuring the content of scenarios and action descriptions in specifications. Thus, there are different case frames (Ohnishi, 1996) for different problem frames, but far fewer case frames than there are systems or problem domains. Scenario in this section is different from the scenario in Section 5.3.3, although we use the same term.

5.4.3.1 Scenario description language

We have developed a scenario description language based on the case grammar (Ohnishi, 2001-b). We specify a sequence of events as a scenario with this language. Each of events has just one verb, and each of verbs has its own case structure (Ohnishi, 1996). Verbs and their own case structures are depend on problem domains, but roles of cases are independent of problem domains. There exist several roles, such as agent, object, recipient, instrument, source, and so on. An entity doesn't have one of these as a type, but plays these as roles. For example, a person may be a user w.r.t. requesting services of the system, but a consumer of its outputs.

We provide action frames as shown in Table 5.3. Each action has its own case structure like Ohnishi's case frame (Ohnishi, 1996). Both action frames and case structures depend on a domain.

Table 5.3 Action frames

Action	Cases
Transform/derive/compute	agent, source, goal, object
Collate/compare	agent, source, object
Report on/ monitor/sense	agent, source, operation, instrument
Query/ask	agent, object, goal
Command/control	agent, recipient, operation, instrument
Suggest/request	agent, object, goal
Create/ edit	agent, source, goal
Feed back	agent, recipient, operation, instrument
Inform/ notify	agent, object, goal, instrument
Allocate/ schedule/ seize	agent, object, goal
Cancel/ relinquish	agent, recipient, operation, instrument
Exist/locate	agent, status

Move/transfer/input/output	agent, source, goal
React	agent, operation, instrument

5.4.3.2 Scheduling and control example: elevator scheduler

In this section, we consider a scenario of elevator control problem (Lift Problem, 1987). First we show an informal scenario description, and then show a formal representation based on the action frame model.

[Scenario from passenger's viewpoint]

The following scenario is an informal scenario from passenger's viewpoint.

We regard a scenario as a description of a sequence of events.

1) *Passenger is at floor 2.*
2) *Passenger decides to go to Floor 3.*
3) *Passenger enters a car.*
4) *Passenger pushes Button[3]*
5) *Car feeds back to Passenger the status(going to 3) by lighting both the Dir[UP] and Button[3] on*
6) *Car starts UP*
7) *Car arrives at 3 and stops*
8) *Car feeds back to Passenger the status(at 3) by lighting DIR[UP] and Button[3] off and ringing the bell*
9) *Doors open*
10) *Passenger gets off the car*

Each of events can be transformed into a formal representation based on the action frame model. For example, the 5th event can be transformed into the following table.

Table 5.4 Internal representations of events

concept ·	agent	recipient	operation	instrument
feed back	car	Passenger	light ON	Dir[UP]
feed back	car	Passenger	light ON	Button[3]

[Scenario from system's viewpoint]

In this section, we specify a scenario of the same system behavior from system's viewpoint.

1) *Car is now at floor 2*
2) *Doors are opening*
3) *Button[3] in the car is pushed by passenger*
4) *LiftCtrl(Lift Controller) sets the goal of car to go to floor 3.*
5) *Car lights Dir[UP] and Button[3] on*
6) *LiftCtrl sets motor UP.*
7) *Dir[UP] lights on, and Button[3] lights on*

8) LiftCtrl closes the doors
9) Doors are closing
10) LiftCtrl starts motor UP
11) LiftCtrl gets the state of Car at floor 3
12) LiftCtrl stops motor
13) LiftCtrl lights Dir[UP] and Button[3] off
14) LiftCtrl rings the bell[3]
15) *LiftCtrl opens the doors*

5.4.3.3 Transformation rules between scenarios from different viewpoints

Table 5.5 Scenario transformations with rules.

from Passenger's view	from System's view	Rule#
1. Passenger B is at floor 1		
2. Passenger decides to go to Floor 4		
3. He calls the car by pressing UP[1]	1. LiftCtrl senses passenger B's pressing the Button UP[1] 2. LiftCtrl set the goal to floor 1	1 1
4. LiftCtrl feeds back the status by lighting the UP[1] button ON	3. LiftCtrl lights on UP[1].	3
5. Car arrives and stops	4. Car stops the motor 5. Motor stops	4 4
6. Car feeds back to him the status by ringing the bell	6. Car rings the bell	3
7. Doors open	7. LiftCtrl open the doors 8. Doors open	4 4
8. He enters		
9. He pushes Button[4]	9. LiftCtrl senses passenger B's pressing the Button[4] 10. LiftCtrl set the goal to floor 4	1 1
10. Car feeds back the status to him	11. Car lights Dir[UP] and Button[4] on	3
11. Doors close	12. LiftCtrl close the doors 13. Doors close	4 4
12. Car starts up	14. LiftCtrl starts motor UP 15. Motor starts UP	4 4
13. Car arrives at floor4 and stops	16. LiftCtrl stops motor 17. Motor stops	4
14. Car feeds back the status to him	18. Car lights Button[4] and Dir[UP] OFF 19. Car rings the bell	3 3
15. Doors open	20. LiftCtrl open the doors 21. Doors open	4 4
16. He gets off		

By comparing the scenario from user's viewpoint with the scenario from system's viewpoint, we found four rules of transforming between these two scenarios as shown below.

1) ***Command by user to system corresponds to Sense by system***
2) ***Command by user to system should be fed back to user***
3) ***Feed back to user corresponds to command by system***
4) ***Reaction for passenger corresponds to pair of command by LiftCtrl and reaction by its subsystem***

The 1st, 2nd and 3rd rules can be applied to most of interactive system. The 4th rule depends on the elevator system problem domain. These rules can be regarded as domain models for transformation rules between scenarios from different viewpoints.

We provide a new scenario from passenger's viewpoint and transformed it into a scenario from system's viewpoint with these 4 rules as shown in Table 5.5.

As shown in the above table, we can derive a new scenario specified from system's viewpoint from a scenario specified from passenger's viewpoint. 21 events in the new scenario can be automatically transformed from the events in the old scenario with the transformation rules.

We have developed an action frame base scenario description language and transformation rules between scenarios from different viewpoints. With these rules, we can get a new scenario from system's viewpoint by transforming a scenario from user's viewpoint. Through our examples, we found the scenario transformation with action frame enables to improve several quality of scenario.

1) System developer does not need to describe scenarios from several different viewpoints. If he specifies a scenario from one viewpoint, then he can get a scenario from another viewpoint with transformation rules. This contributes to improve the productivity of scenarios.
2) If the system developer specifies two scenarios from different viewpoints, he can check the lack of events and can check the consistency of scenarios by comparing a scenario with a transformed scenario. This contributes to improve the completeness and the consistency of scenarios.

Scenario description language, action frame, and transformation rules between scenarios are dependent on problem domain. We have to validate the ideas more thoroughly by applying to several different problem domains.

5.4.3 Other topics

5.4.3.1 Patterns and Requirements Engineering

In (Fowler, 1997) and (Nakatani, 1999), software patterns in requirements analysis of business application domain are illustrated. These patterns include hierarchy of organization, accountability, resource allocation, and so on. Good patterns for the reusability and the generalization are described (Tomoeda, 1999).

Patterns specified with object diagrams, sequence diagrams, and state charts can be regarded just domain models. There exist several researches for using analysis patterns to define software requirements in a certain problem domain in (Gaska, 1998), (Lecouce, 1998) and (Johnson, 1996).

By using such patterns, requirements analysis, especially object-oriented analysis will be performed more effectively and we can get a requirements specification of good quality.

5.4.3.2 Requirements Engineering based on Reusing Software Processes

In DAM, software including models, software specifications, documents, and source codes are regarded as domain models. Reuse of the domain models is mainly focused in DAM, but reuse of software processes should be considered, because the reuse of software process enables to support the requirements definition phase. For example, process of project management, process of organization of software development team, process of adopting software development method and environments in a successful software development in a certain problem domain can be reused in another software development in the same domain. In other words, in case of successful requirements definition in a certain domain,

1) process of project management
2) process of organization of software requirements definition team
3) process of requirements definition

are reusable to support another requirements definition of the same problem domain.

5.5 CONCLUSION

Here, both Requirements Engineering based on DAM and DAM based on Requirements Engineering are described with examples. The author hopes that the discussions at the DAM WG and this paper will contribute to advance researches for the members of the Working Group of DAM, IPS, Japan.

ACKNOWLEDGEMENT

The author would like to thank to Professor Kiyoshi Itoh at Sophia University, Japan for his kind invitation to the Working Group of DAM and this Domain Oriented Systems Development (DOSD) book. The author also thanks to Associate Professor Colin Potts at Georgia Institute of Technology (GA Tech), Atlanta, U.S.A. for his useful comments about the research of scenario transformation when the author visited GA Tech as a visiting scientist. The author wishes to thank heartily to all the members of both RE WG and DAM WG, IPS Japan for their valuable comments.

BIBLIOGRAPHY

Boehm, B.W., 1981, Software Engineering Economics, (Prentice-Hall, Inc.).

Davis, A.M., 1993, Software Requirements, Objects, Functions, & States, (Prentice-Hall, Inc.).

Fowler, M., 1997, Analysis Patterns: Reusable Object Models, (Addison-Wesley Publishing Co. Inc.).

Gaska, M. T., Gause, D. C., 1998, An Approach for Cross-Discipline Requirements Engineering Process Patterns, In *3rd International Conference on Requirements Engineering (ICRE)*, (IEEE Computer Society), pp.182-189.

Itoh, K., Hirota, T., Kumagai, S., Yoshida, H. ed., 1998, Domain Oriented Systems Development: Principles and Approaches, *the series of Advanced Information Processing Technology*, Vol.1, (Gordon and Breach Science Publishers).

Lecouce, R., Mellish, C., Robertson, D., 1998, A Framework for Requirements Elicitation through Mixed-Initiative Dialog, In. *3rd ICRE,* (IEEE Computer Society), pp.190-196.

Jackson, M., 1995, Problems and requirement, In *2nd International Symposium on Requirements Engineering (ISRE)*, (IEEE Computer Society), pp.2-8.

Johnson, R. E., Nakamura, H., Nakayama, Y., Yoshida, K., 1996, Pattern and Framework (in Japanese), (Kyoritsu-Shuppan).

Jones, C., 1996, Applied Software Measurement, Assuring Productivity and Quality, 2nd ed., (McGraw-Hill Co.).

Nakatani, T., Aoyama, M., Sato, K. eds., 1999, Software Pattern (in Japanese), *bit*, (Kyoritsu Shuppan).

Ohnishi, A., 1996, Software Requirements Specification Database Based on Requirements Frame Model, In *2nd ICRE*, (IEEE Computer Society), pp.221-228.

Ohnishi, A., 1999, VRDL: A Visual Software Requirements Language, In *Trans. Society for Design and Process Science (SDPS)*, Vol.3, No.3, (SDPS), pp.43-52.

Ohnishi, A., 2001-a, Report of Requirements Engineering Working Group (in Japanese), In *SIG Software Engineering*, Vol.2001, No.31, (Information Processing Society of Japan (IPSJ)), pp. 127-134.

Ohnishi, A., Potts, C., 2001-b, Grounding Scenarios in Frame-Based Action Semantics, In *7th International Workshop on Requirements Engineering*: (Foundations of Software Quality (REFSQ'01)), (Essener Informatik Beitrage), pp.177-182

OMG Object Management Group, 2000, Unified Modeling Language Specification, (OMG).

Sommerville, I., 1995, Software Engineering, 5th ed., (Addison-Wesley).

Tomoeda, A., Tamura, Y., Sakamoto, N., 1999, Reuse of Pattern Architecture with use of Architecture-oriented Analysis (in Japanese), In *Object oriented Symposium'99* (oo'99), IPSJ Symposium Series, Vol.99, No.9, (IPSJ), pp.51-58.

"Lift Problem," 1987, In *4th International Workshop on Software Specification and Design* (IWSSD), (IEEE Computer Society), p.x.

CHAPTER 6

The Know-It-All Project: A Case Study in Framework Development and Evolution

Greg Butler

Ling Chen

Xuede Chen

Ashraf Gaffar

Jinmiao Li

Lugang Xu

Concordia University

ABSTRACT

The Know-It-All Project is investigating methodologies for the development, application, and evolution of frameworks. A concrete framework for database management systems is being developed as a case study for the methodology research. The methodology revolves around a set of models for the domain, the functionality, the architecture, the design, and the code. These models reflect the common and variable features of the domain. There is a set of alignment maps between models that provides traceability while preserving commonality and variability.

From a database perspective, our aim is to provide support for all data models, integrated and heterogeneous databases, and eventually support for incomplete and uncertain data. The current prototype covers the relational data model. Our short-term goal is to add deductive databases and graph databases, thus providing diagrammatic queries to support applications in bioinformatics.

6.1 INTRODUCTION

Despite the dramatic improvement on design techniques and tools supports, it is still difficult and error-prone to design and implement complex software. Systematic reuse using domain engineering, product lines, or frameworks has been successful in delivering the economic benefits of reuse. An object-oriented framework is a concrete realization in source code of a domain-specific software architecture. While frameworks are often cited as an example of the reuse of requirements, architecture, design, and code, there is a prevailing conception that the only concrete model used for frameworks is the source code itself.

The Know-It-All project, which has been underway at Concordia University since 1997, has three sets of aims:

1. to research methodologies and models for framework development, application, and evolution;

2. to develop a framework for database management systems, supporting a variety of data models of data and knowledge, the integration of different paradigms, and heterogeneous databases; and

3. to apply the Know-It-All framework to advanced database applications for bioinformatics. Bioinformatics is a good application domain since it involves the management and analysis of large volumes of diverse data, with ever increasing volume and variety.

The research on software methodology in an academic setting needs a concrete case study in order to evaluate the methodology and models; the framework case study needs to be evaluated by creating real applications. For us the database framework is a case study to validate our methodology, while the applications to bioinformatics allow us to validate the database framework. A side-benefit of this is that, along the way, the research in software technology leads to a platform for research in database technology, which in turns leads to advances in bioinformatics and genomics.

Methodologies for the development of a framework have been suggested that use domain analysis, software evolution, and design patterns. However, identifying the required flexibility for the family of applications and designing mechanisms that provide this flexibility is the problem. Furthermore, the evolution of the framework must be considered, especially as all frameworks seem to mature from initial versions through to a stable platform.

Refactoring of source code is one approach suggested for the development and evolution of frameworks. Refactoring is a behavior-preserving program transformation, such as moving methods up the class hierarchy. Our work attempts to show how a framework can be described by a set of models, not just source code, and how refactoring as a concept can be broadened to apply to each of these models. Our view of refactoring includes the underlying motivation for the restructuring performed during the refactoring. This motivation could be expressed as an issue, rationale, or force (or sets of these). We call our approach cascaded refactoring. It is the process of sequentially refactoring each model, one after the other, where the restructuring of the previous model provides the motivation for the refactoring of a model.

6.2 BACKGROUND

Frameworks (Fayad *et al.*, 1999) offer a concrete realization of a product line. A framework is an architecture, plus an implementation, plus documentation that capture the intended use of the framework for building applications. A framework provides a highly effective mechanism for software reuse within an application domain. The framework captures the features that are common across the product line. In return for relinquishing some design authority, the developer can build a new application faster by hooking to the framework just the code that is unique to the new application.

A domain is an area of knowledge that is scoped to maximize the satisfaction of the requirements of its stakeholders; includes a set of concepts and terminology understood by practitioners in that area; and includes the knowledge of how to build software systems in that area.

The basic properties of a framework are (a) modularity, since abstract classes with stable interfaces, encapsulate and localize change in "hot spots", the points of planned variability; (b) reusability of analysis, design, and code; (c) extensibility by providing explicit hot spots or "hooks" for planned variability; and (d) inversion of control since the framework calls the custom code.

6.2.1 Terminology

There is confusion with terminology since different communities using different words–– feature, hook, hot spot — to describe essentially the same concept: a variable aspect of the system.

- A feature is any aspect of a system used to characterize it to a stakeholder. This includes both common and variable aspects, but the variable features are the most interesting.
- A hook is a point in the framework that is meant to be adapted in some way, such as by filling in parameters or creating subclasses.
- A hot spot is a variable aspect of a framework domain, whereas a fixed aspect is called a frozen spot.

6.2.2 Framework Development Approaches

We can classify the approaches to framework development as

- example-driven, bottom-up, incremental development;
- top-down domain engineering;
- hot spot generalization;
- use case-driven; and
- hybrid, which combines one or more of the above.

These classifications are not mutually exclusive.

The classic development approach in the object-oriented community (Roberts and Johnson, 1997) is example-driven, bottom-up, and incremental. A framework is refactored as new requirements are encountered as the applications are developed one by one.

Domain engineering (STARS, 1996) often invests more effort up-front in the modeling of the domain and the specification of an architecture for the product line.

Their understanding of the domain is more complete, and is captured in range of models, including the context, scope, taxonomy, data dictionary, feature model, domain specific software architecture, descriptions of major functionality as use cases and algorithms, and exemplar systems in the domain.

Hot spot generalization (Pree, 1995; Schmid, 1997) identifies the points of flexibility — the hot spots — required in the product line; classifies the type of flexibility required (using patterns or meta-patterns); and then assigns a subsystem with a facade class within the design to implement the hot spot.

The use case-driven approaches extend the OOSE methodology of Jacobson from simple applications to product families. They include Catalysis, RSEB (Jacobson *et al.*, 1997), and Feature RSEB (Griss *et al.*, 1998).

6.3 OUR METHODOLOGY AND MODELS

Our methodology for framework development and evolution is a hybrid approach. It combines the modeling aspects of the top-down domain engineering approaches, and the iterative, refactoring approaches of the bottom-up object-oriented community. The basic philosophies are

Framework development is framework evolution.
Evolution = refactoring + extension.

The methodology weaves together steps for partial domain engineering to better understand the domain and how to evolve the current working set of partial models, and steps of system refactoring and extension.

The methodology stresses traceability between the models. In this respect it follows FeatureRSEB (Griss *et al.*, 1998) and FORM (Kang *et al.*, 1998). In the context of a framework or product line, the methodology must also stress the commonality-variability distinction. Butler (1999) defines an alignment of models to be a traceability mapping that is consistent with the mapping of common and variable features.

While the working set of partial models is incomplete, and hence some mappings for traceability or alignment will be partial maps, we still desire a consistent, coherent, aligned set of models and maps.

Our models are

- M_f, the feature model;
- M_u, the use case model;
- M_a, the architectural model;
- M_d, the design or class hierarchy;
- M_s, the source code;

and also a test plan and test cases as part of the documentation.

Feature Model A feature is any aspect of a software system that is used to characterize the system to the users, customers, or developers. A feature model is a collection of all the features and their relationships. For a product line or framework, a feature model classifies each feature as mandatory, optional, or alternative in order to express the commonality and variability across the applications in the product line. An application is specified by its feature set, which is the set of all features provided by the application. A feature set is a subset of the set of all features, and the feature set satisfies the constraints amongst features imposed by the model.

There are several views of structure for a feature model. The basic concept is a finite set F of the features. The usual relationship of interest is the subfeature relationship that defines a hierarchy of features. There is also a dependency relationship between features, and the actual dependency may be specified using a constraint. One can consider the subfeature relationship as a special kind of dependency.

Structure is also provided by two classifications. The usual classification into mandatory, optional, and alternative can be considered as a map variability from F to the set {mandatory, optional, alternative}. The FORM methodology adds a classification of features into capability features, domain technology features, operational environment features, and implementation features. This can be considered as a map kind from F to the set {capability, technology, environment, implementation}.

Use Case Model Use cases capture the functionality of a system, called the target system, as it is meant to behave in a given environment called the host system. A use case describes how a group of external entities, called actors, make use of the system under consideration. The use they make is modeled by the passing of signals or information between the actors and the system.

A use case is a description of a cohesive set of dialogues that the primary actor initiates with a system. The dialogues are cohesive in the sense that they are related to the same task, or form part of the same transaction. Cohesiveness is often determined by having a goal in common for the tasks, or by having a common responsibility.

The high-level view of use cases provides a generalization and inclusion relationship between use cases. Abstract use cases may occur in the model to express commonality amongst use cases, even if the abstract use case has no concrete scenario; its (leaf) children in the generalization hierarchy will have concrete scenarios. The inclusion relationship connects a task with a subtask represented as a use case.

Extension points and variation points in a use case model allow the variability in functionality to be depicted.

A use case encompasses a collection of scenarios, since there may be several ways in which an actor can (successfully or unsuccessfully) attempt a task. Potts *et al.* (1994) defines a scenario as "a description of one or more end-to-end transactions involving the required system and its environment". Scenarios may be classified as follows. The main scenario describes the usual way in which the task is successfully performed. Typically, in the main scenario, the simplest sequence of interactions is described, and it is assumed that all steps execute successfully. A variant scenario describes another way of using the system where it is assumed that all steps execute successfully. An exceptional scenario describes a scenario where exceptional or error conditions may arise. It may be possible to recover from the exceptions and therefore successfully complete the task — this is called a recovery scenario — or it may not be possible to recover — this is called a failure scenario.

A scenario may be described in several ways, from a simple narrative text description, to numbered steps indicating the subject-action-object triples, to basic Message Sequence Charts (MSCs) (Regnell *et al.*, 1996) or UML sequence and collaboration diagrams.

A use case may also be described in terms of episodes. Each episode represents a subtask, or the parts of the dialogues in the scenario that perform the subtask. A high-level Message Sequence Charts (MSCs) can be used to depict a use case and its episodes utilizing operators for sequence, alternation, iteration, parallel execution, and exceptions.

Use cases may be classified into different kinds, such as business use case, requirements use case, analysis use case, system use case, and design use case. These reflect the perspective of the writer, as well as the level of detail and the kind of target system.

Architectural Model Software architectures provide an abstract description of the organizational and structural decisions that are evident in a software system. The development of an architecture requires the decomposition of system into subsystems, the distribution of control and responsibility, and the development of the components and their connections or means of communication. General principles of information hiding, such as the use of modules, layers, and abstract machines (API's) help manage the complexity of the task.

An architectural style (Shaw and Garlan, 1996) may be described in terms of

- a vocabulary of the basic design elements (components and connector types),
- a set of configuration rules which constrain how components and connectors may be configured,
- a semantic interpretation which defines when suitably configured designs have a well-defined meaning as an architecture, and
- analyses that may be performed on well-defined designs.

From the perspective of UML and RUP (Kruchten, 1995), the description comprises four diagrams plus the use cases/scenarios. The diagrams cover the design view, the process view, the implementation view, and the deployment view. The design view presents the subsystems, their interfaces, the dependency between subsystems, and nesting of subsystems.

It is possible to take a use case view of a subsystem, where the subsystem is regarded as the target system with the host system consisting of all other subsystems in the architecture. Then a subsystem's services are described in the use case model for the subsystem, and accessed through its interfaces.

It is also possible to take a class view of a subsystem, where one identifies a subsystem with a facade class. One identifies the subsystem interfaces with the class methods.

At present we model architecture using layers, subsystems, and interfaces using the UML notation. So far that has been adequate. However, in the future, we plan to investigate the use of the Siemans approach (Hofmeister *et al.*, 1999) to architectural modeling.

Design Model The design is modeled in the classic object-oriented style separating static structure and dynamic behavior with the class diagrams, interaction diagrams, and state diagrams of UML. Design patterns usage is encouraged.

Trace Maps The alignment maps used are

- T_{fu}, the trace map from the capability features M_f to the use case model M_u;

- T_{fa}, the trace map from the operational environment features M_f to the architectural model Ma;
- T_{fd}, the trace map from the domain technology features M_f to the design M_d;
- T_{fs}, the trace map from the implementation features M_f to the source code M_s;
- T_{ua}, the trace map from the use cases M_u to the architecture M_a;
- T_{ud}, the trace map from the use cases M_u to the design M_d;
- T_{ad}, the trace map from the architecture M_a to the design M_d; and
- T_{ds}, the trace map from the design M_d to the source code M_s.

6.3.1 Framework Evolution by Cascaded Refactoring

Evolution is naturally viewed as refactoring followed by extension in those bottom-up methodologies for framework development that are centered on refactoring. The new structure of the system and the refactorings to effect the re-structuring are chosen with the required extension in mind. However, there is little literature on this connection between refactoring and extension. We also say very little on this topic. A useful reference on extension of object-oriented systems is Mätzel and Bischofberger (1997).

Refactoring (Fowler, 1999) is a behavior-preserving program transformation that automatically updates an application's design and underlying source code. Primitive refactorings perform simple edits such as adding new classes, creating instance variables, and moving instance variables up the class hierarchy. Compositions of refactorings can create abstract classes, capture aggregation and components, and even install design patterns (Tokuda, 1999).

Our methodology extends the notion of refactoring that has been applied to source code, and to design in the form of class diagrams, to other models of software systems. The methodology relates the set of refactorings across the set of models, through change impact analysis using the trace maps. This process, called cascaded refactoring, is a series of refactorings of the models. The impact of the refactorings for a model M_x is translated via the trace maps that have domain M_x to determine constraints on the refactorings of models M_y. The methodology is unique in several ways:

- to view refactoring as an issue-driven activity;
- to document the rationale of an application of a refactoring as a triple: intent of restructuring, choice of refactoring(s), and impact of the restructuring; and
- the notion of cascaded refactoring, where the restructuring of one model determines constraints on the restructuring of other models (via the traceability and alignment maps).

We still need to investigate how the choice of refactorings is influenced by the desired extensions of the framework. Butler and Xu (2001) made contributions to

- refactoring of feature models;

- refactoring of use case models;
- preliminary work on refactoring of architectures; and
- clarify the "behavior" that is preserved by these refactorings.

6.3.2 Documenting Frameworks

The application of a framework to create a product line is hampered by three issues:

1. the learning curve for a framework is too steep;
2. the framework itself must be stable before application development can be undertaken on a large scale; and
3. there is a lack of experience across the maturity lifecycle of a framework — customizing a white-box framework is very different from applying a generator to build an application.

Our research (Butler and Dénommée, 1999; Butler *et al.*, 1998) has clarified many of the issues, and recommended a set of documentation to assist application developers.

- An overview of the framework setting a context for the domain and the variability in the framework.
- A graded set of example applications illustrating the customization of features.
- A cookbook of recipes of how to customize each feature, organized in a spiral fashion from easiest/most common to most advanced/least common, and using the examples to keep the presentation concrete.

We view documentation as a key step in framework development, since the ability to write clear documentation that explains how application developers should use the framework means that the concepts of the design are clear and that the steps required for customization have been clearly thought out. Hence, documentation verifies that the framework is easy to use, and this is the overriding goal of framework design.

6.4 THE CASE STUDY FOR DOMAIN OF DATABASES

Know-It-All is an object-oriented framework for database management systems. It is written in C++, with some Java for user interfaces, and XML for communication of data between the C++ framework and the Java tools. The user interfaces will provide a full range of query mechanisms, from icons for canned queries, to forms, to textual queries in set comprehension languages, and diagrammatic queries.

Our domain analysis considered common database textbooks and research literature, commercial and public-domain systems, and research prototypes. The domain is huge, as indicated by a summary in Table 6.1, however, our initial scope can be seen in the top-level feature model Figure 6.1.

Table 6.1 Some Dimensions of the Database Domain

Dimension	Features
Data Model	none, flat file, relation, deductive, graph, object, statistical, time-sequence, string, text, structured text, multimedia, spatial, GIS, constraint
Functionality	browse, query, bulk load, update, transaction, trigger, stored procedure, integrity constraint, workflow
DBA Functions	configure, monitor, tune, evolve, auto-configure, auto-tune
Feature	concurrency, recovery, security, client/server, distributed, heterogeneous
Performance	data clustering, compression, encryption, index, query optimization, view materialization, caching
Scale	#user, #transactions, #databases, #processors

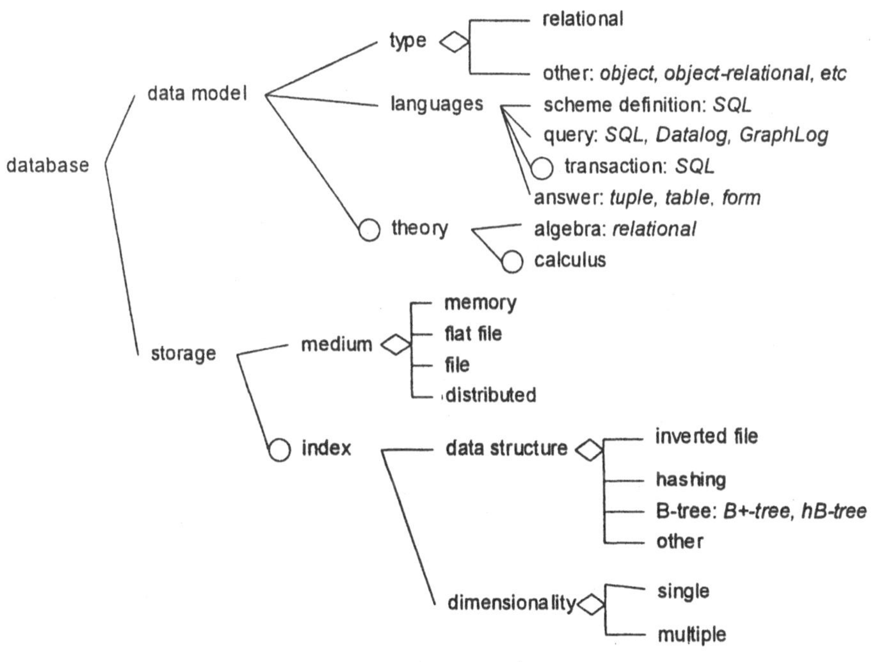

Figure 6.1 Top-level Feature Model of the Know-It-All Framework

The use case model Figure 6.2 not only shows the major functionality of the system, but also the context of the system and the roles and interaction between users, application developers and database administrator.

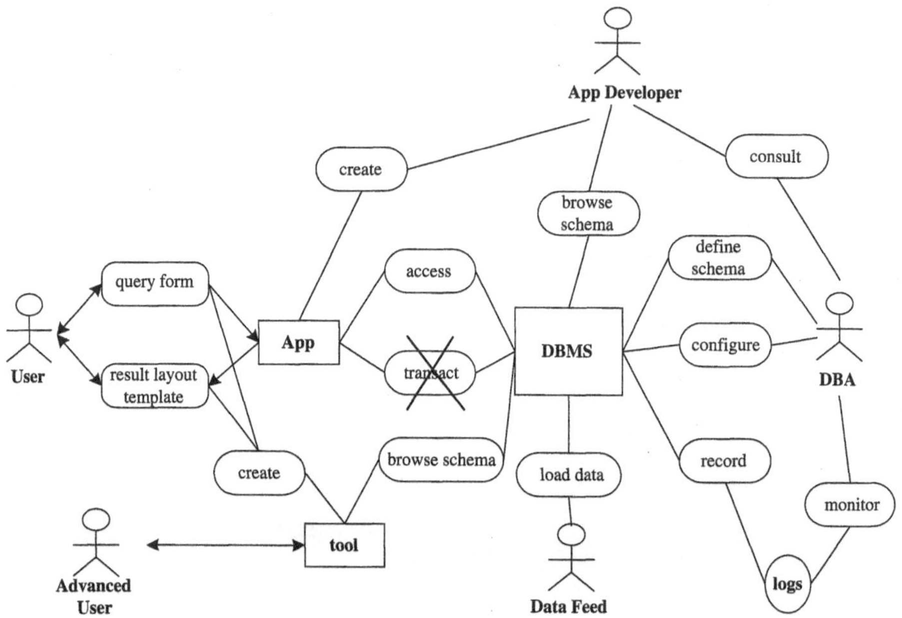

Figure 6.2 Use Case Model of the Know-It-All Framework

Know-It-All is designed with scientific databases in mind, and does not provide for transactions. Instead, it provides a data feed mechanism for bulk or incremental data loads. The prime concern is querying of existing data. The framework provides a generic infrastructure for database management systems and allows them to support a range of data models (relational, object, object-relational, etc) where the data model itself, and its constituents for query language, query optimizing, indexing, and storage have clearly defined roles. A database in Know-It-All is seen as a series of layers, each of which provides the same interface. The usual breakdown of responsibilities into physical, logical, conceptual, and view layers is followed by Know-It-All. However, a database, as seen by the end-user, allows views of views, and mappings of object conceptual models to relational conceptual models. Eventually, Know-It-All will incorporate composite databases (such as integrated or heterogeneous databases) and make no distinction between simple and composite databases.

Each layer in Know-It-All is basically a translator between its client layer and its supplier layer(s), as shown in Figure 6.3. A layer provides a mechanism to decompose or translate queries, and a mechanism to reconstruct answers (for example, an execution plan for relational algebra expressions). The translation is done with the aid of the schema, and produces both the translated query, and the mechanism to reconstruct answers. The layer architecture is adapted from Mackinnon *et al.* (1998)'s architecture for heterogeneous databases, while the reconstruction mechanism is designed as an iterator (which is a tree of iterators) that returns the next result (Graefe, 1996)

Our prototype implements the relational data model. For support of diagrammatic queries, it is a priority to support deductive and graph databases. For general needs in genomics, we need to support object databases. This will allow us to support spatial, temporal, and image databases. There are now also algebras for object-relational databases, so they will also be supported.

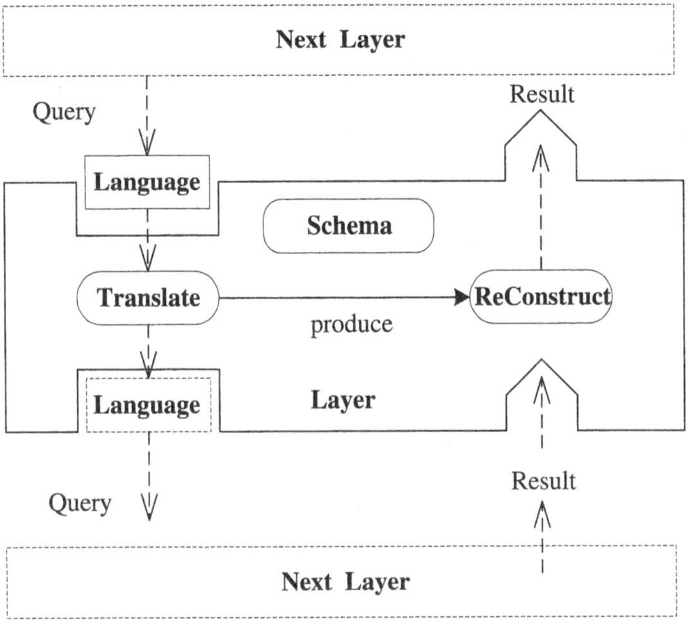

Figure 6.3 Basic Building Block

6.4.1 Subproject: Query Optimization Framework

The Know-It-All framework contains a subframework for query optimization, and a subframework for indexing techniques. The optimization framework is based on the broadly applicable OPT++ (Kabra and DeWitt, 1999). Figure 6.4 shows the overall class diagram. In this figure, components of this framework are represented as packages and only the top-level classes in each component are shown.

The Search Strategy component encapsulates the system control flow. It implements different search approaches that are used to perform the search in the search space. At any instance of time, only one search approach dominates. It also encapsulates the cost model and implements the cost estimation framework. It contains the logical query plans and physical query plans that have been built, and it performs the cost evaluation on these plans and prunes out the sub-optimal ones. The Search Strategy component maintains an aggregation reference to the Algebra component because each logical query plan is represented by applying an operator (in the Algebra component) to its root node, and each physical query plan is represented by applying an algorithm (in the Algebra component) to its root node. The major classes in the Search Strategy component include: the QUERYOPTIMIZERFACADE, the SEARCHSTRATEGY, the SEARCHTREE, the OPERATORTREE, and the ALGORITHMTREE. The class QUERYOPTIMIZERFACADE is the facade of this system. It simplifies the use of the system. The SEARCHSTRATEGY is an abstract class and provides interfaces for all search strategy approaches that are used in query optimization. The SEARCHTREE represents the search tree that is used to explore the optimal plan. It implements the search strategy interfaces. The OPERATORTREE represents a logical query plan of a query. It is an algebraic expression that represents the particular operations to be performed on data and the necessary constraints regarding order of operations. The ALGORITHMTREE represents a physical query plan of a query. It is a tree of algorithms that specifies the particular order of operations and the algorithm used to implement each operation.

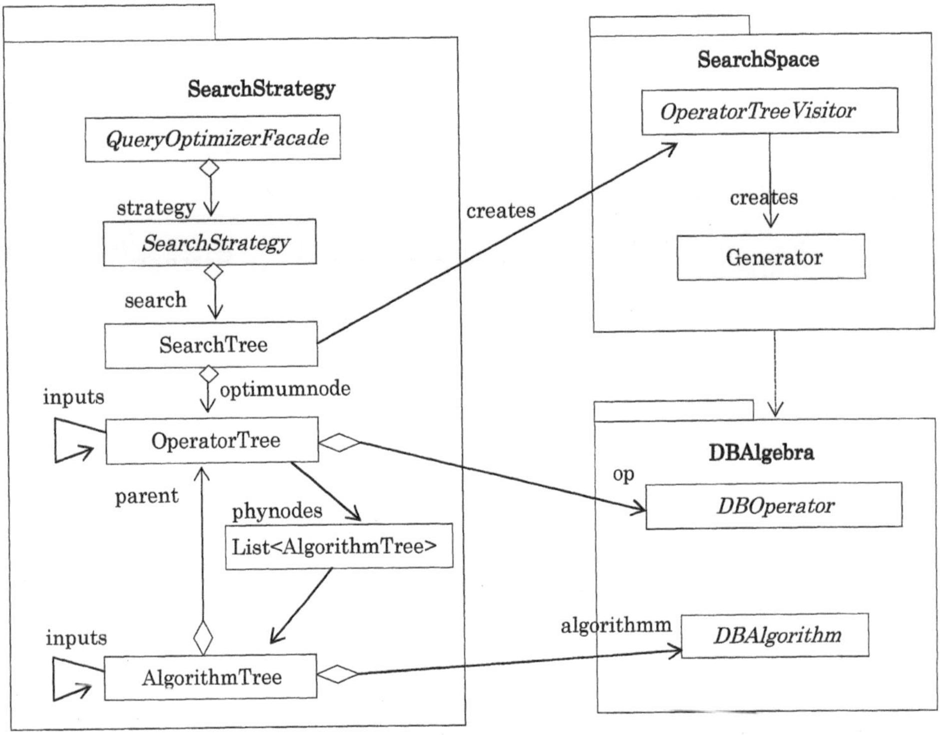

Figure 6.4 Overall Class Diagram

The Algebra component defines the logical operators (operations on the database) and the physical operators (execution algorithms for these logical operators) in the database. There are two major classes in the Algebra component: the DBOPERATOR and the DBALGORITHM. The former defines the interfaces for all possible logical operators in the database system. The latter defines the interfaces for all possible execution algorithms for these logical operators. They are all abstract classes and depend on the optimizer-implementor to define the actual database algebra. The definitions of the interfaces for the logical and physical operators should consider easy extension of possible operations that are performed on them.

The Search Space component defines what the search space is. It defines the operations on the Algebra. The effects of executing the operations in the Search Space component are:

- A tree is transformed to another tree. The new tree becomes bigger as a result of expansion or these two trees are equivalent as a result of some equivalent transformations.

- A logical query plan is converted to a physical query plan.

Conceptually, there are two search spaces. One is the logical search space. It is a space of logical query plans and decides how one logical query plan derives from another. The other is the physical search space. It is a space of physical query plans and decides how one physical query plan derives from another. The conversion from the logical query plan to the physical query plan can be deemed as an operation on the logical search space. There are two major classes in the Search Space component. One is the OPERATORTREEVISITOR and the other is the Generator. The OPERATORTREEVISITOR is an abstract class and defines all operations performed on the Algebra component. It serves as a bridge between the Algebra component and the Search Space component. The Generator is not a class defined in this framework. In fact, there are three generator classes, each is defined for one of the following operations: tree expansion; tree transformation; or conversion of a logical query plan to a physical query plan.

6.4.2 Subproject: Tree Index Framework

The indexing subframework is based on GiST (Hellerstein *et al.*, 1995), which covers tree-based indexes, including multi-dimensional trees and similarity-based retrieval. We have re-designed GiST to follow the design of STL containers in C++, to follow good object-oriented design practices, and use design patterns. In the future, the indexing subframework will be extended to cover inverted files and hashing techniques.

6.4.3 Subproject: Diagrammatic Query Interface

Data management, access, and mining are at the heart of bioinformatics. While relational databases are the accepted standard within industry, there has been considerable research into deductive databases and graph databases to extend the capabilities of relational databases. Deductive databases allow a view, called the intensional database, to be defined using logical rules, and allow logical queries against the view. Since the rules allow recursive definitions, the resulting expressive power of the query language is greater than ordinary relational databases. Graph query languages are even more expressive, while having the very important property of a visual representation.

Consens et al. (1994) at the University of Toronto developed the GraphLog graph query language based on hygraphs, and a visual interface, called Hy+, for expressing queries and browsing their results. GraphLog is a graph query language extending Datalog and negation. The language has recursion, usually as transitive closure, and has path expressions. Path expressions are similar to regular expressions. A path expression can refer to a primitive relation, or construct more complex path expressions using the operators of negation, inverse, concatenation, alternation, kleene closure (*), or transitive closure (+).GraphLog is more expressive than Datalog (and SQL). Hy+ is a visual Smalltalk environment for composing GraphLog queries and viewing the results of the query. In the diagrams, nodes represent entities, while labels add identifiers and attributes to nodes. Edges represent relations and edge labels are path expressions. Blobs are nodes containing other nodes and represent a relation on containing and contained nodes. The underlying data structure of GraphLog and Hy+ is a hygraph.

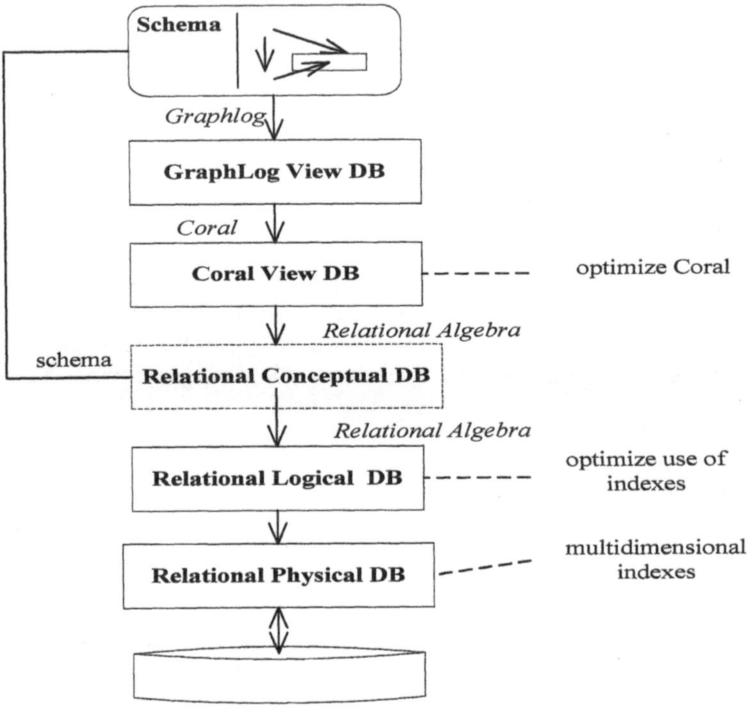

Figure 6.5 Hy+ and GraphLog in the Know-It-All Framework

The query interface for Hy+ imports the database schema, together with iconic descriptions for each class of entities and relations. The icons for relations are various forms of lines and arrows. A query, or a view definition, is constructed by selecting icons and composing a hygraph. A hygraph is basically a graph augmented with blobs. A blob relates a containing node with a set of contained nodes, and can be viewed as providing an aggregation or subset relation useful for abstraction of complex graphs.

The interface provides a window for the definition of views. A view may define a property of an entity, as a relation between the entity class and a "ground" symbol. A view may define a relation in terms of pre-existing concepts in the database and the views. A view may also define a blob in terms of pre-existing concepts in the database and the views. The icons for the defined property and relation arrows, and for blobs, appear in the schema window along with the icons for the database entities and relations. Closure of a relation is depicted as a dashed arrow of the same color as the underlying relation. A negation of a relation has a cross through the arrow.

The interface provides a window for the definition of queries. While composing a query, it is possible to have open the windows containing the definitions of the views that are used by the query. This, together with the schema window, guides the user to make well-formed queries.

The translation of GraphLog diagrams into logic programming notations such as Coral (Ramakrishnan *et al.*, 1994) or Datalog is straightforward. The hard part is ensuring efficient processing of the translated query. There has been no experimentation with Coral optimization strategies, and one really needs multidimensional indexes of the underlying relations to support efficient processing of the deductive programs.

Hygraphs are also the means for presentation of results of queries. The presentation is controlled initially by specifying a layout algorithm for the hygraph in a layoutGraphLog window, and by specifying which entities or relations to hide in a hideGraphLog window. As before, the use of icons and colors aids comprehension of the hygraphs. Hy+ also provided the ability to zoom in for more detail. The window for the presentation of results displays the hygraphs of the results, and provides tools for abstraction by collapsing blobs. It is very natural in Hy+ to refine queries by simply regarding the current query as a view definition, and then composing a query (a refinement) against the database and using the new view. Of course, there is the potential to use the set of results as a materialized view.

While the initial Java version of the Hy+ interface will be implemented using Coral, we will eventually incorporate it as part of our framework. The first step is to incorporate deductive databases, in the form of Coral, into the Know-It-All framework for databases. This will be done by regarding the deductive database as a VIEWDB subclass defining an intensional database view of a relational database. See Figure 6.5. The second step is to incorporate graph databases directly into the Know-It-All framework, also as a VIEWDB subclass of either relational, object, or object-relational databases. Then we will incorporate recent techniques for the processing of path queries.

6.5 APPLICATIONS TO BIOINFORMATICS

This is not the place for a long expose on genomics and bioinformatics, so we content ourselves with directing the reader to some of our other publications. Butler *et al.* (2000) is an introduction to bioinformatics and its need for information technology together with our perspective on future directions for the technology. Butler (1999) is more specific to database technology for pathway databases, while Letovsky (1999) covers the major database systems in bioinformatics.

6.6 CONCLUSION

The Know-It-All project is investigating methodologies for the development, application, and evolution of object-oriented frameworks. Although the research is being done in an academic setting, we are attempting to provide solid evidence of advances in modeling and methodology through the development of a case study framework for the domain of databases, and to apply this framework to real scientific applications.

The essence of the methodology — cascaded refactoring — has been formulated. It is an issue-driven process of restructuring and extension of a coherent set of (partial) models of the domain and the framework. The models underlay a use-case driven view of system development (as in RSEB), augmented by a layered feature model (as in FORM) to better model commonality and variability.

Many details of the methodology are still being investigated. This includes the choice of architectural model or models, and precise definitions of the trace maps for each pair of models. However, most of the unknowns about the methodology lie in cataloguing refactorings for the various models, and relating desired extensions of the framework to the choice of refactorings.

A subproject to develop modeling and refactoring tools for the use case models is underway, and we hope to initiate a similar subproject for the architectural models soon. These should throw much light on the gaps in our methodology.

Our methodology is still weak in treating extensibility, for example, how the planned extensions affect the choice of refactorings, and how to perform the extensions. We will face these issues as our case study matures and we observe its evolution.

The Know-It-All framework case study has completed a partial domain analysis and prototyped its architectural design in C++ for the relational data model. The query optimization subframework has been completed, resulting in an implementation, models, and documentation on its use. The tree index framework will be completed in 2001.

The GraphLog interface, as a standalone system, will be finished in the near future, and over the longer term both Coral and GraphLog will be incorporated as data models in the Know-It-All framework.

Potential future work on the Know-It-All framework includes a subframework for physical storage, a subframework for hash indexes, and a subframework for inverted file indexes. The range of data models included in the framework will also be extended.

At present our bioinformatics platform is using the mySQL database for its data repository. In the short term, we plan to create a mySQL-compatible instance of the Know-It-All framework with the relational data model, and to use this as a mirror of the data repository. This will allow us to use the enhanced querying available in the framework, and allow us to delay the implementation of transaction management in Know-It-All.

BIBLIOGRAPHY

Butler, G., 1999, Developing frameworks by aligning requirements, design, and code. In *9th Workshop on Software Reuse* (WISR-9).

Butler, G., 1999, Database technology for pathways, In *Workshop on Computation of Biochemical Pathways and Genetic Networks*, edited by Bornberg-Bauer, E., de Beuckelaer, A., Kummer, U., Rost U., (Logos Verlag, Berlin), ISBN 3-89722-093-8, pp. 89-95.

Butler, G., Bornberg-Bauer, E., Grahne, G., Kurfess, F., Lam, C., Paquet, J., Rojas, I., Shinghal, R., L., Tao, R., Tsang A., 2000, The BioIT projects: Internet, database and software technology applied to bioinformatics. In SSGRR', Suola Superiore G. Reiss Romoli SpA, Coppoto, Italy. URL http://www.ssgrr.it/en/ssgrr2000/proceedings.htm

Butler, G. and Dénommée P., Documenting frameworks. In [8], pp.495-504.

Butler, G., Grogono P. and Khendek, F., 1998, A reuse case perspective on documenting frameworks. In *Asia-Pacific Software Engineering Conference*, pp. 94-101.

Butler, G. and Xu, L., 2001, Cascaded refactoring for framework evolution. In *2001 Symposium on Software Reusability*, (ACM Press), pp. 51-57.

Consens, M.P., Eigler, F.Ch., Hasan, M.Z., Mendelzon, A.O., Noik, E.G., Ryman, A.G. and Vista, D., 1994, Architecture and applications of the Hy+ visualization system. *IBM Systems Journal*, **33**, pp. 458-476.

Fayad, M.E., Schmidt, D.C. and Johnson, R.E., 1999, *Building Application Frameworks: Object-Oriented Foundations of Framework Design*, (Wiley).

Fowler, M., 1999, *Refactoring: Improving the Design of Existing Code*, (Addison-Wesley).

Graefe, G., 1996, Iterators, schedulers, and distributed-memory parallelism. *Software - Practice and Experience,* **26**, pp. 427-452.

Griss, M. L., Favaro, J. and d'Alessandro, M., 1998, Integrating feature modeling with the RSEB. In *Fifth International Conference on Software Reuse*, (IEEE Computer Society), pp. 76-85.

Hellerstein, J.M., Naughton, J.F. and Pfeffer, A., 1995, Generalized search trees for database systems, In *VLDB'1995*, pp. 562-573.

Jacobson, I., Griss, M. and Jonsson. P., 1997, *Software Reuse: Architecture, Process and Organization for Business Success*, (Addison-Wesley).

Hofmeister, C., Nord, R. and Soni, D., 1999, *Applied Software Architecture*, (Addison-Wesley).

Kabra N., and DeWitt, D.J., 1999, OPT++: An object-oriented implementation for extensible database query optimization. *VLDB Journal,* **8** pp. 55-78.

Kang, K. C., Kim, S., Lee, J., Kim, K., Shin, E. and Huh. M., 1998, FORM: A feature-oriented reuse method with domain-specific reference architectures. *Annals of Software Engineering*, **5**, pp. 143-168.

Kruchten, P., 1995, The 4+1 view model of architecture. *IEEE Software*, pp. 42-50.

Letovsky, S.I., 1999, *Bioinformatics: Databases and Systems*, (Boston, Kluwer Academic Publishers).

Mätzel, K.-U. and Bischofberger, W., 1997, Designing object systems for evolution. *Theory and Practice of Object Systems*, **3**, pp. 265-283.

Mackinnon, L.M., Marwick, D.H. and Williams, M.H., 1998, A model for query decomposition and answer construction in heterogeneous distributed database systems. *Journal of Intelligent Information Systems*, **11**, pp. 69-87.

Potts, C., Takahashi, K. and Anton, A., 1994, Inquiry-based requirements analysis. *IEEE Software*, pp. 21-32.

Pree, W., 1995, *Design Patterns for Object-Oriented Software Development*, (Addison-Wesley).

Ramakrishnan, R., Srivastava, D., Sudarshan, S. and Seshadri, P., 1994, The CORAL deductive system. *VLDB Journal,* **3**, pp. 161-210.

Regnell, B., Andersson, M. and Bergstrand, J., 1996, A hierarchical use case model with graphical representation, In *Proceedings of Second International Symposium on Engineering of Computer-Based Systems*, (IEEE Computer Society Press), pp. 270-277.

Roberts, D. and Johnson, R., 1997, Patterns for evolving frameworks, In *Pattern Languages of Program Design 3*, (Addison-Wesley), pp. 471-486.

Schmid, H.A., 1997, Systematic framework design by generalization. *Communications of the ACM*, **40**, pp. 48-51.

Shaw, M. and Garlan, D., 1996, *Software Architecture: Perspectives on an Emerging Discipline*, (Prentice Hall).

STARS, 1996, Organization domain modeling (ODM) guidebook — version 2.0. Technical report, (Lockheed Martin Tactical Defense Systems).

Tokuda, L., 1999, Evolving object-oriented architectures with refactorings, In *Proceedings of ASE-99: The 14th Conference on Automated Software Engineering*, (IEEE CS Press).

CHAPTER 7

Rapid Knowledge Base Development for Product Configuration Systems using the Unified Modeling Language

Alexander Felfernig

Gerhard Friedrich

Dietmar Jannach

Markus Zanker

Institut für Wirtschaftsinformatik und Anwendungssysteme,

Universität Klagenfurt

ABSTRACT

Knowledge-based product configuration systems play an important role in modern business strategies. These systems support the sales representative or the technical engineer to cope with the complexity of configurable products, the huge number of available variants, and different restrictions on allowed product constellations. Nowadays, typical knowledge-based configuration systems are not well integrated into standard software development processes but use proprietary knowledge representation formalisms which are not understandable for domain experts. In this chapter we show how the Unified Modeling Language (UML) can be applied as domain-oriented notation for the design of configuration knowledge bases. The provided modeling concepts can be used for comprehensible knowledge acquisition and are given precise semantics, such that an automatic translation to executable configuration knowledge bases is feasible. Following a discussion on these modeling concepts we show how their expressiveness can be enhanced by using the Object Constraint Language (OCL) and how the construction of complex configuration models can be supported.

7.1 INTRODUCTION

In today's rapidly changing, globalizing markets traditional mass production paradigms appear anachronistic. Mass production is increasingly replaced by customer-individual production of highly variant products. Companies are forced to diversify their product spectrum in order to be able to fulfill the individual needs of customers. The additional costs for offering customer-individual products must be minimized in order to be able to provide customizable products in a competitive fashion. “Mass-customization” (Anderson, 1997) appeared as a new paradigm representing the trend towards the production of highly variant products under mass production speed and pricing conditions. This paradigm imposes increasing demands on the development and maintenance of software supporting sales and production of highly variant products. This software must be able to handle rapidly changing, complex constraints on the products and on the corresponding processes supporting quotation, order processing, production, delivery, and maintenance. In particular knowledge-based configuration systems (configurators) are increasingly applied for supporting processes related to highly variant products.

There exists a variety of application domains for product configuration systems. One well known application domain is the telecommunication industry, where telecommunication systems support switching functionality for enterprise-wide telephone services. Furthermore, these systems also provide additional services like voice-over-ip, messaging services, ISDN connections, video-telephony, or video-conferencing. In this context configurators are used for calculating bill-of-materials representing the configuration of a switching system. The configuration of audiovisual equipment, automobiles, computer systems, trucks, airplanes, modular furniture (kitchens etc.), industrial products (e.g. valves, actuators, or controls), or light control systems are further application areas for knowledge-based configuration.

Effective application of configuration technology implies a number of improvements. The configuration system automatically checks the requirements imposed by the customer with regard to given marketing constraints, technical constraints, and constraints concerning the production process, which reduces the response time to given customer requests. Automated configuration avoids errors in the quotation and order processing phase. Consequently, time consuming reconfigurations of non-realizable orders are avoided and the time between equipment sales and delivery/installation of the product is decreased.

Configuration systems can significantly contribute to the improvement of processes related to highly variant products and are of strategic importance for enterprises. However, the development and maintenance of these systems is facing a set of challenges, which must be tackled in order to allow an effective application of configuration technology. First, the complexity of the task requires the sophisticated knowledge of experts. This knowledge must be effectively acquired and translated into a corresponding configuration knowledge base. Second, configurator development time is short and strictly limited since the development of the product and the product configuration system have to be done in parallel. Finally, the configuration knowledge base has to be adapted continuously because of changing sets of available components and configuration constraints.

In the following we show, how the knowledge acquisition task for configuration knowledge bases can be effectively supported by applying the Unified Modeling Language (UML) (Rumbaugh and Booch, 1998) as domain-specific representation language for building configuration knowledge bases. We employ UML, since this language is widely applied in industrial software development as a standard design language supporting the software development process starting with the requirement analysis phase proceeding until the implementation phase. UML has a built-in constraint language which supports a formal definition of constraints on models that are designed using the graphical concepts of UML class diagrams. UML is extensible for domain-specific purposes, i.e. the semantics of the basic modeling concepts of the language can be further refined in order to be able to provide domain-specific modeling concepts, which allow a more intuitive construction of the corresponding models. Finally, we have made excellent experiences in using UML designs for validation by technical experts.

7.2 CONFIGURATION TASK

In order to avoid time-consuming development of specific designs for each customer, customizable products are designed following a building block principle where basic parts can be configured into different sorts of assemblies. A configuration task can be characterized through a set of available component types, a description of their properties, namely attributes and possible attribute values, connection points (also called ports), and constraints on legal combinations of those constitutive elements. Given some customer requirements, the result of computing a configuration is a set of components, corresponding attribute valuations, and connections satisfying all constraints and customer requirements.

In Felfernig *et al.* (2000-c) an application-independent first-order logic-based definition of a configuration task is given. In this work a configuration problem is defined as a logic theory based on two sets of logical sentences.

1. Domain description (DD). Configurations are composed from components corresponding to component definitions in a component catalog – these definitions are described by a set of *types*. Attributes of component types are described by the function *attributes*, the domains of those attributes are described by the function *dom*. Finally, ports of component types are described by the function *ports*. The domain of these functions are sets of constants describing the corresponding attributes, attribute domains, and ports. These sets including a set of constraints on legal combinations of components, connections, and value instantiations are assumed to be contained in the set DD.

2. System Requirements Specification (SRS). Most configuration problems incorporate some kind of additional requirements (e.g. customer requirements) which describe additional constraints respectively initial components which must be part of the configuration. These constraints are contained in SRS.

In Felfernig *et al.* (2000-c) a configuration problem is defined as follows.

Definition (Configuration Problem). *In general a configuration problem is described by a triple (DD,SRS,CONL), where DD and SRS are sets of logical sentences and CONL is a set of predicate symbols. DD represents the domain description, and SRS specifies the particular system requirements which define an individual configuration problem instance. A configuration CONF is described by a set of positive ground literals whose predicate symbols are in the set of CONL.*

Note that - depending on the configuration domain – the set of predicates in CONL is extensible if needed. For example, a predicate could be defined specifying that a certain port of a component must be left unconnected. A consistent configuration is defined as follows in Felfernig *et al.* (2000-c).

Definition (Consistent Configuration). Given a configuration problem (DD,SRS,CONL), a configuration CONF is consistent iff DD $\cup$ SRS $\cup$ CONF is satisfiable.

The intuitive definition of a consistent configuration allows determining the validity of partial configurations, but does not require completeness of the configuration (e.g. a partial configuration only containing a floppy disk is a consistent configuration).

It is necessary, that a configuration explicitly includes all needed components as well as their connections and attributes. Therefore we need to introduce an explicit formula for each predicate symbol in CONL (completeness axioms) to guarantee this completeness property. In order to stay within first-order logic, this property is modeled by first order formulae. A set of completeness axioms (Felfernig, *et al.*, 2000-c) assures the completeness of CONF. Such a complete configuration is denoted as CCONF.

Definition (Valid and Irreducible Configuration). *Let (DD, SRS, CONL) be a configuration problem. A configuration CONF is valid iff DD $\cup$ SRS $\cup$ CCONF is satisfiable. CONF is irreducible if there exists no other valid configuration $CONF_{sub}$ such that $CONF_{sub} \subset CONF$.*

As mentioned in Friedrich and Stumptner (1999) the above definitions represent a high-level view of the language developed and used in the COCOS configuration project (Fleischanderl, *et al.*, 1998), which used as representation formalism a generative constraint satisfaction scheme that can be defined by a direct mapping from the consistency-based semantics sketched in this section.

7.3 BUILDING CONFIGURATION KNOWLEDGE BASES IN UML

The unavailability of methods for handling the increasing size and complexity of software systems triggered the "software crisis" in traditional software development. Similarly, knowledge-based systems development experienced increasing challenges triggered by the size and complexity of the knowledge bases. R1/XCON (Barker, *et al.*, 1989) can be seen as an representative example for the complexity of knowledge base maintenance. In 1989 the knowledge base had about 31.000 components and about 17.500 rules (engineering, manufacturing, and marketing rules) with a changing rate about 40% (rule additions, deletions, and modifications) per year.

In order to tackle these challenges, a number of methods and tools supporting the expert system development process were proposed (e.g. Schreiber, *et al.*(1994)). Using these approaches, the development of knowledge-based systems was supported and improved by the provision of diagrammatic notations and formal description languages. Similar approaches were also developed in Software Engineering research in order to effectively handle the implementation of software systems and to support a structured development process for these systems. These approaches have progressed from functional decomposition and data-driven techniques and methods (e.g. Cameron (1989)) to object-oriented analysis and design techniques and methods such as Rumbaugh, *et al.*(1991), Booch, *et al.*(1994). Moving from data-driven and functional approaches, research focused on the development of object-oriented analysis and design approaches, which are nowadays also accepted in industrial software development processes. These approaches reduce the "semantic gap" between the considered universe of discourse and the abstract model and increase the understandability and maintainability of the resulting models (Jacobson, *et al.*(1992)).

The Unified Modeling Language (UML) (Rumbaugh and Booch, 1998) is the result of an integration of the object-oriented approaches of Booch, *et al.*(1994), Jacobson, *et al.*,(1992), and Rumbaugh, *et al.*(1991), which is well established in industrial software development processes. UML is applicable throughout the whole software development process from the requirements analysis phase to the implementation phase providing different notations. One of the major advantages of UML is the principle of extensibility of the basic meta-model. Following this principle the basic modeling concepts defined in the UML meta-model can be further extended in order to support the definition of domain-specific modeling concepts for specific application areas. The set of domain-specific modeling concepts including constraints on their syntactical usage are defined in an UML *profile*[1].

State-of-the-art approaches for developing software systems provide methods and techniques which support an effective construction of those systems. However, the transformation of informal requirements into an implemented system is still a challenge for the development of knowledge-based systems as well as for traditional software development. In the following we show how to integrate the development of configuration systems into the standard software development process by using UML as a knowledge acquisition front-end which allows the automatic generation of executable configuration knowledge bases.

7.3.1 Example: UML Configuration Model

For the following discussions the simple UML configuration model shown in Figure 7.1 will serve as working example.

This model represents the generic product structure, i.e. all possible variants of a configurable car. The basic structure of the product is modeled using classes, generalization, and aggregation. The set of possible products is restricted through a set of constraints which are related to technical restrictions, economic factors, and restrictions according to the production process. The used modeling concepts can be seen as an ontology in the sense of Chandrasekaran, *et al.* (1999), i.e. ontologies are theories about the sorts of objects, properties of objects, and relations between objects that are possible in a specified domain of knowledge. This ontology represents an integration of different basic configuration paradigms (Soininen, *et al.* (1998)). Component types (represented through the stereotype "ComponentType") represent the constituent parts of a configuration model; *car*, *engine*, or *gearing* are examples for such component types (see Figure 7.1). Component types are organized in a part-of hierarchy, where aggregations between component types can be either composite or shared aggregations in the sense of the UML semantics (Unified Modeling Language Specification, 1999).

Furthermore, resources and functions are further basic concepts which are represented as stereotyped classes in a UML configuration model[2]. Parts of a configuration problem can be seen as a resource balancing task, where some of the component types produce some resources and others are the consumers. In many cases some parts of the configuration model contain information which are relevant for the customer, e.g. not every screw of a car must be presented to the customer. Functions are used to represent exactly those parts of the configuration model. Similar to component types, functions can be organized using partonomies and taxonomies.

Beside the structural information, a configuration model also contains a set of constraints on the correct usage of the components and functions within a configuration. Such constraints can also be expressed graphically.

In Figure 7.1 the existence of an *automatic gearing* requires the existence of an *otto engine* within the final configuration. Furthermore, a *4-wheel gearing* must not be used in combination with a *diesel engine* within the final configuration. These constraints can be used to express basic restrictions on the configuration model, but must not be seen as sufficient for expressing all kinds of constraints on a configuration model.

---

[1] For building product models in UML we have defined an UML configuration profile.

[2] These concepts are not used in the example model.

In order to express constraints on configuration models which cannot be expressed graphically, we employ the Object Constraint Language (OCL)[3].

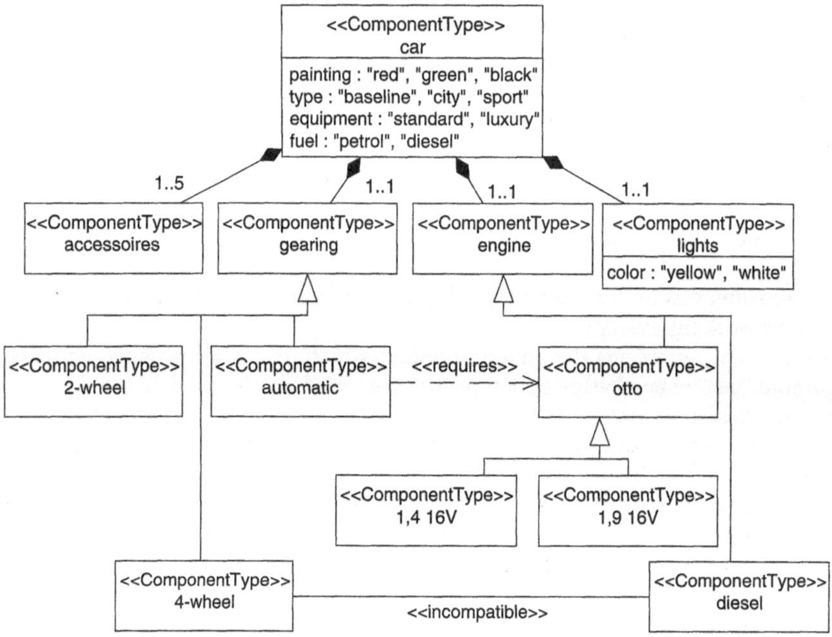

Figure 7.1 A simple UML Configuration Model

7.3.2 Introducing Domain-specific Modeling Concepts in UML

In order to introduce a modeling concept, into the configuration profile, the following steps are necessary.

1. Define the new concept (a new stereotype or tagged value[4]) and state the well-formedness rules for its correct usage within the model. Well-formedness rules are expressed as OCL constraints on the UML metamodel. In order to restrict the usage of the stereotyped dependency "requires" (see the requires relation between the *automatic gearing* and the *otto engine* in Figure 7.1) to the connection of two component types, the following constraint can be formulated.

   context dependency inv:
   self.stereotype.name="requires" implies
   self.client.stereotype.name="ComponentType" and

[3] The application of OCL will be discussed in Section 7.4.

[4] *Stereotypes* and *tagged values* are basic mechanisms provided by UML for defining restrictions on the UML meta-model.

self.supplier.stereotype.name="ComponentType".

2. Define the semantics of the concept for the configuration domain by stating the facts and constraints induced to the logic theory when using the concept. These translation rules are discussed in detail in Felfernig, *et al.* (2000-a,b).

In order to state additional constraints on the product model which cannot be expressed graphically, the Object Constraint Language (OCL) can be employed. In Section 7.4 we sketch, how OCL can be used to define additional constraints on configuration models.

7.3.3 Derivation of Configuration Knowledge Bases

In order to support the automatic construction of a configuration knowledge base from the conceptual UML model, a clear definition of the semantics of the employed modeling concepts has to be provided (see Felfernig, *et al.* (2000-a,b)). This is achieved by defining translation rules from the conceptual configuration model to a logical theory (Felfernig, *et al.*, 2000-c). Such a theory includes the description of a part library and constraints on those parts. The result of the translation is a set of first-order logical sentences that form a domain description which can be used by a configuration system which is based on the component-port representation (Mittal and Frayman, 1989) and (Friedrich and Stumptner, 1999).

In the following we do not discuss the translation rules in detail, but give a couple of examples for the result of applying the translation rules. A discussion of the rules for translating UML class diagrams and OCL constraints into configuration knowledge bases can be found in Felfernig, *et al.*(2000-a,b).

The configuration knowledge base which corresponds to the UML configuration model of Figure 7.1 is the following.

types={car, accessoires, gearing, engine, lights, 2-wheel, automatic, 4-wheel, otto, 1-4-16V, 1-9-16V, diesel}.
attributes(car)={painting, type, equipment, fuel}.
attributes(lights)={color}.
dom(car, painting)={red, green, black}.
dom(car, type)={baseline, city, sport}.
dom(car, equipment)={standard, luxury}.
dom(car, fuel)={petrol, diesel}.
dom(lights, color)={yellow, white}.
ports(car)={accessoires1,...,accessoires5, gearing, engine, lights}.
ports(accessoires)={car}.
ports(gearing)={car}.
ports(engine)={car}.
ports(lights)={car}.

The component types of Figure 7.1 are translated into a set of *types* with corresponding *attributes* and attribute *domains (dom)*. Furthermore, the partof relationships are translated into *port* constants which are used to express connections between components in a configuration.

The constraints defined on the configuration model of Figure 7.1 are translated into the logic representation of a configuration problem as follows. It is assumed, that

CONL={type/2, conn/4, val/3}, where *type(ID, a)* indicates that *ID* is of type *a*, *conn(ID1, p1, ID2, p2)* indicates that component *ID1* is connected via port *p1* with *ID2* and *ID2* is connected with *ID1* via port *p2*. Furthermore, *val(ID, a, c)* indicates that the attribute *a* of *ID* has the value *c*.

> *An "automatic" gearing requires the existence of an "otto" engine:*
> *type(ID1, automatic) ∧ type(ID2, car) ∧ conn(ID1, car, ID2, gearing) ⇒ ∃ ID3 type(ID3, otto) ∧ conn(ID3, car, ID2, engine).*
>
> *A "4-wheel" gearing is incompatible with a "diesel" engine:*
> *type(ID1, 4-wheel) ∧ type(ID2, car) ∧ type(ID3, diesel) ∧ conn(ID1, car, ID2, gearing) ∧ conn(ID3, car, ID2, engine) ⇒ false.*

The form of logical sentences is restricted to a subset of range-restricted[5] first-order-logic with set-extension, the variables [6] are all-quantified if not explicitly mentioned. In order to assure decidability, the term-depth is restricted to a fixed number. For implementation purposes the logical sentences can be regarded as instantiation schemes (Friedrich and Stumptner, 1999) for the translation into other representation formalisms, such as a Generative Constraint Satisfaction Problem (GCSP (Stumptner and Haselböck, 1993)) representation.

A configuration result consists of a set of positive ground literals whose predicate symbols are in CONL. Assuming that *CONL={type/2, conn/4, val/3}*, the following set of ground literals could represent a configuration result CONF.

> *CONF={type(id1, car). type(id2, accessoires), type(id3, automatic). type(id4, 1_4_16V). type(id5, lights). val(id1, painting, red). val(id1, type, city). val(id1, equipment, standard). val(id1, fuel, petrol). val(id5, color, white). conn(id1, accessoires, id2, car). conn(id1, gearing, id3, car). conn(id1, engine, id4, car). conn(id1, lights, id5, car)}.*

7.4 EXTENDING CONFIGURATION MODELS USING OCL

The Object Constraint Language (OCL (Warmer and Kleppe, 1999)) is an expression language based on first-order logic which enables the definition of constraints on object-oriented models. OCL was strongly influenced by the development of Syntropy (Cook and Daniels, 1994), an object-oriented modeling language which combined the diagrammatic representation concepts of OMT (Rumbaugh, *et al.*, 1991) with the formality of Z (Diller, 1994). OCL complements the graphical (semiformal) notation of UML by providing a precise vocabulary for expressing constraints on graphical models. Since UML is a wide-spread modeling language in industry, OCL itself establishes an increasingly important role in the field of formal specification languages. However, the definition of the OCL semantics is based on a proposed syntax (represented as context-free grammar) and additional textual descriptions and examples. Although this is a quite intuitive approach for demonstration purposes, a corresponding formal definition is needed. In order to provide a semantics for OCL constraints in product configuration, we

---

[5] Every variable of the consequence part of the clause is also contained in the antecedent part.

[6] Variables are starting with an upper case letter.

propose a set of translation rules that transform those constraints into the logic representation discussed in the previous sections[7].

A constraint imposed to the configuration model of Figure 7.1 could state that a *red car* requires *yellow lights*. This constraint can be formulated as follows in OCL.

context car inv:
car.painting="red" implies car.lights.color="yellow"

This constraint is built from the following parts:

- *Context:* A context describes for which class the constraint has to hold. In the above example the constraint has to hold for all *car* instances.
- *Navigation expression:* The navigation expression *car.lights* results in the lights component connected with the car instance. The name *"lights"* is the default name of the association from the class *car* to the class *lights* since no association name is specified here (Unified Modeling Language Specification, 1999).
- *Attribute access:* access to attributes of objects. In the above example, the attributes *painting* and *color* are accessed.
- *Operator "=":* In the above example this operator is used for comparing the attribute values of *painting* and *color* with corresponding constants ("red", "yellow").

Such an OCL constraint can be translated into a closed logical formula following an implication schema, where the right hand side (RHS) contains the all-quantified variables which correspond to the result of the evaluations of navigation expressions and the corresponding operations. The left hand side (LHS) contains the corresponding logical consequence, i.e. the evaluation results of the right hand side connected with the logical and relational operators.

The example OCL constraint on the car configuration model is translated as follows into the logic representation of a configuration problem.

type(ID1, car) ∧ *ResultSet1={ID1}* ∧ *connected_set_a(ResultSet1, painting, ResultSet2)* ∧ *connected_set(ResultSet1, lights, ResultSet3)* ∧ *connected_set_a(ResultSet3, color, ResultSet4)* ∧ *Val1* ∈ *ResultSet2* ∧ *Val2* ∈ *ResultSet4* ⇒ *(Val1=red* ⇒ *Val2=yellow)*.

The predicates *connected_set* and *connected_set_a* are used to express a component navigation respectively an attribute access in a given OCL navigation expression, e.g. *connected_set(ResultSet1, lights, ResultSet3)* expresses the navigation from a *car* instance to the connected *light* instance, *connected_set_a(ResultSet1, painting, ResultSet2)* expresses an access to the attribute *painting* of the class *car*.

7.5 STRUCTURING CONFIGURATION KNOWLEDGE BASES

The application of UML for building configuration models is motivated by the wide-spread use of the language, the high degree of understandability, the extendibility for domain-specific purposes, and the availability of a built-in constraint language which

[7] A discussion of the translation rules can be found in Felfernig, *et al.*(2000-b).

allows the definition of complex constraints. However, when modeling highly variant products, that offer the customer a vast amount of interdependent options, expressiveness restrictions of a single (partitioned) diagram are approached. The diagrammatic depiction becomes harder to maintain and understand as the number of elements and constraints depicted in the diagram increases. In order to improve the understandability and maintainability of UML configuration models, we introduce the notion of contextual diagrams (see Felfernig and Zanker (2000), Felfernig, *et al.* (2000-d)), which provide a means for structuring constraints in a configuration model.

7.5.1 UML Packages

Before introducing contextual diagrams, let us have a look at the basic structuring mechanisms provided in UML. A package is simply defined in UML as a grouping of model elements. Packages themselves may be nested within other packages. It is therefore straightforward to partition a diagram into packages under the premise of high cohesion among the elements in the same package and low coupling between different packages.

The strictly defined hierarchical package structure can not only be used to partition the model knowledge, but as well to structure the solving process of the underlying configuration system. In some cases parts of a configuration problem can be solved more or less independently. This property can be used to structure the solving process, which leads to smaller problem sizes and less computational complexity in the solving process. The partitioning of configuration problems for allowing efficient calculations of solutions for large configuration problems is discussed in Fleischanderl and Haselböck (1993). The configuration of switching systems is one of the most complex application areas for knowledge-based configuration. As mentioned in Fleischanderl, *et al.* (1998), configurations representing switching systems can become large and complex. Fleischanderl, *et al.* (1998) describe an example system with about 43.000 components, 215.000 attributes and 120.000 ports. In order to allow an efficient calculation of solutions, the problem domain must be separated into different independent sub-problems – exactly this property was exploited in Fleischanderl and Haselböck (1993).

7.5.2 Additional Structuring Mechanisms

Parts of the configuration knowledge sometimes apply only in a specific context. For example, depending on the country, different keyboards must be added to a personal computer configuration, furthermore the language of the operating system must be adequately configured. Consequently, the country is a parameter of the configuration model, which determines a set of additionally relevant constraints. If the customer wants to have a certain type of car (e.g. a car of type *baseline*), this selection can also impose further constraints which only apply in the case that this specific type is selected. Figure 7.2 shows the result of adding a set of additional constraints to the configuration model shown in Figure 7.1.

Such constraints can concern different classes sometimes stored in different packages of the configuration model. In real world settings, where the number of constraints and component types is large, the maintenance of models following the strategy shown in Figure 7.2 obviously leads to maintenance problems and is a challenge for the knowledge engineer as well as for the technical expert. For example, if constraints concerning the type of a car must be changed, these constraints are spread all over the configuration model assigned to classes stored in different packages. A closer look at the constraints shows that the constraints can be categorized into groups having the same preconditions. Exactly this property can be used for partitioning the whole model into different views, where each view contains only a restricted set of constraints. We denote such kind of views as contextual diagrams. Constraints concerning a specific context are organized in a central location (contextual diagram), which significantly alleviates the handling of changes. In the example given, constraints concerning the car type *baseline* can be organized in one single contextual diagram.

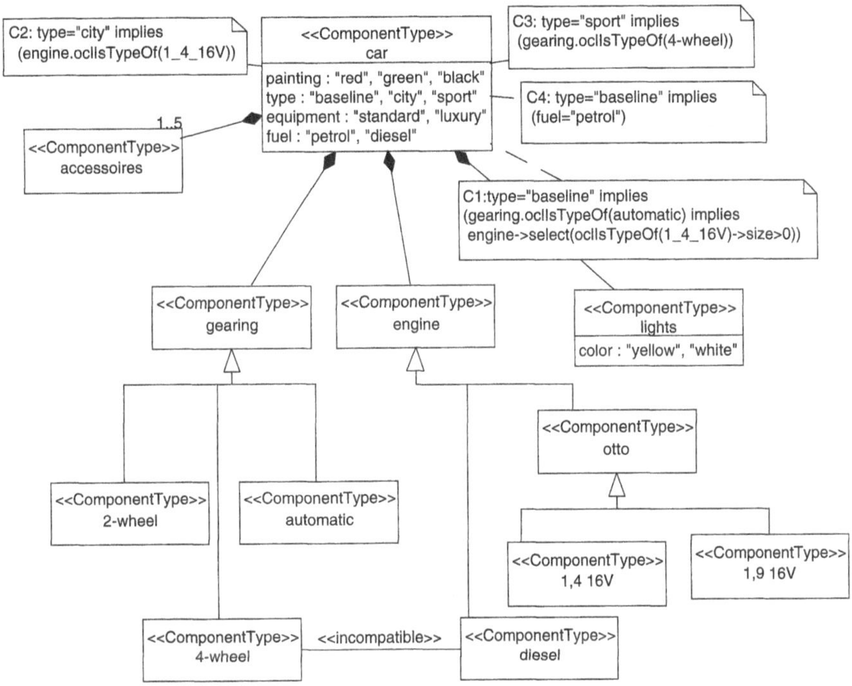

Figure 7.2 Configuration Model with additional constraints

7.5.3 Contextual Diagrams

The notion of context has been discussed in different research areas (Delisle and Schwartz, 1987)' (McCarthy, 1993)' (Siegel, *et al.*, 1992)' (Charlton and Wallace, 2000)' (Theodorakis, *et al.*, 1998)' (Akman and Surav, 1996) using quite different interpretations. Overviews of different interpretations of the notion of context can be found in Charlton and Wallace (2000), Akman and Surav (1996). Informally, a context can be interpreted as a general condition under which an event, action, etc. takes place. The interpretation of the notion of context used in the following is similar to the notion defined in McCarthy (1993) who proposes a formalism *ist(c,p)* in order to define a context *c*, in which the proposition *p* must hold, i.e. *p is true* in *c*. On the graphical level contexts are represented as contextual diagrams. Using contextual diagrams, the graphical representation (GREP) of a configuration knowledge base is organized in a context hierarchy (see Figure 7.3). An important relation defined on contexts is "$\prec$" (see McCarthy (1993)). Such an relation $c_1 \prec c_2$ defines a partial ordering over contexts meaning that the context c_2 contains all the information of the context c_1 and probably more. Except $GREP_{root}$, each contextual diagram $GREP_i$ has an assigned precondition $GREPP_i$, which represents the conditions, under which the constraints defined in $GREP_i$ apply. This precondition can be interpreted as a set of circumstances surrounding the actual configuration process, in which an additional set of constraints must hold. The root model $GREP_{root}$ contains basic structural elements of the configuration model including basic constraints on this model. Contextual diagrams $GREP_i$ can be constructed by imposing further constraints on a copy of an already existing contextual diagram. The result of this construction process is a hierarchy of contextual diagrams (see Figure 7.3), in which each contextual diagram $GREP_i$ either represents the root of the hierarchy ($GREP_{root}$), or is derived from another contextual diagram. Note that all the preconditions of contextual diagrams starting from $GREP_{root}$ along the path to $GREP_i$ are included in $GREPP_i$. For example, the precondition $GREPP_3$ of contextual diagram $GREP_3$ in Figure 7.3 is *c* including the precondition of $GREP_1$, namely *a*, i.e. $a \wedge c$.

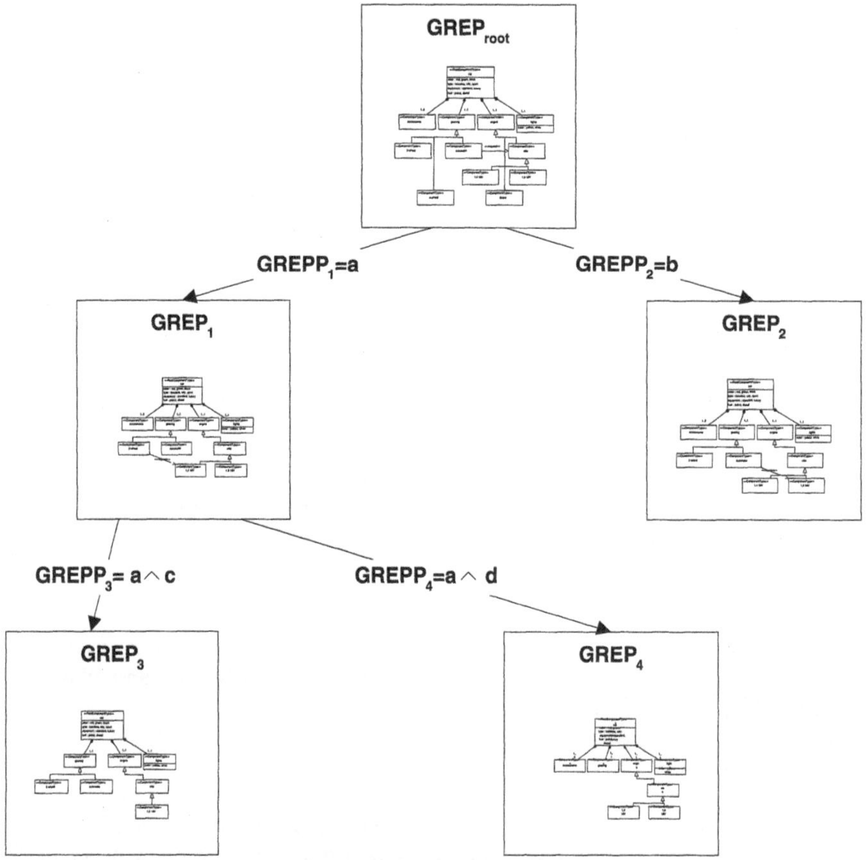

Figure 7.3 A simple Context Hierarchy

We define a context as follows:

Definition (Context). A context c_i is represented by a set of logical sentences $DD_i \subset DD$, where DD represents the domain description. DD_i contains those logical sentences, which solely apply in c_i and are derived from a corresponding contextual diagram $GREP_i$.

Figure 7.4 contains the constraints concerning a car of type *baseline*, which are now included in a separate contextual diagram. $GREPP_{baseline}$ in this diagram is simply the expression *type="baseline"*, which can be formulated as OCL constraint. Note that the knowledge engineer or the domain expert are not forced to enter this precondition textually as OCL constraint, but rather define those restrictions directly in the diagram itself, e.g. by reducing the domain of the corresponding attributes[8]. Constraints, which apply in the actual context can also be entered in a similar way.

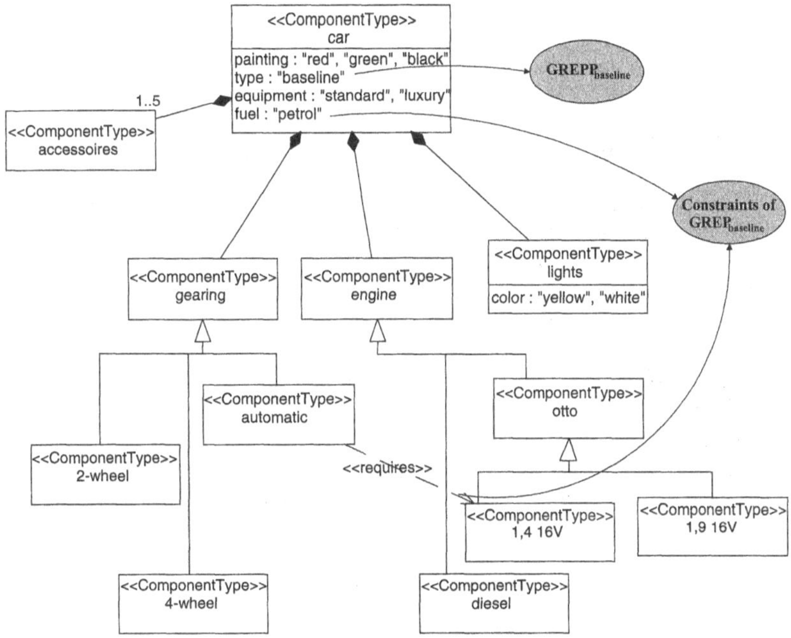

Figure 7.4 Contextual Diagram $GREP_{baseline}$

In order to define a precondition for the contextual diagram $GREP_{baseline}$, the designer of the knowledge base simply reduces the domain of the attribute *type* to *"baseline"* (see Figure 7.4). The constraint *fuel="petrol"* can be entered in a similar way. Finally, a graphical constraint *automatic gearing requires 1,4 16V engine* is defined using a *requires* dependency.

[8] More complex conditions must be formulated as OCL expressions.

The example contextual diagram of Figure 7.4 is translated as follows into the logic representation of a configuration problem, where DD_{root} corresponds to the derived domain description of Section 7.3.3.

An automatic gearing requires a 1_4_16V engine:

type(ID1, car) $\wedge$ *ResultSet1={ID1}* $\wedge$ *connected_set_a(ResultSet1, type, ResultSet2)* $\wedge$ *Val1* $\in$ *ResultSet2* $\wedge$ *Val1 = baseline* $\wedge$ *type(ID2, automatic)* $\wedge$ *conn(ID1, gearing, ID2, car)* $\Rightarrow$ $\exists$*ID3 type(ID3, 1_4_16V)* $\wedge$ *conn(ID3, car, ID1, engine).*

The fuel must be "petrol":

type(ID1, car) $\wedge$ *ResultSet1={ID1}* $\wedge$ *connected_set_a(ResultSet1, type, ResultSet2)* $\wedge$ *Val1* $\in$ *ResultSet2* $\wedge$ *Val1=baseline* $\wedge$ *connected_set_a(ResultSet1, fuel, ResultSet3)* $\wedge$ *Val2* $\in$ *ResultSet3* $\Rightarrow$ *(Val2=petrol).*

7.6 RELATED WORK

The goal of Aldanondo, *et al.* (2000) is to propose modeling concepts enabling a non-computer specialist to describe generic configuration models. Based on a set of requirements concerning different classes of configuration problems a set of modeling concepts is presented which is based on the Dynamic CSP (see Mittal and Falkenhainer (1990)) approach. This approach seems to have its advantages when modeling configuration problems for Dynamic CSP solving, but also seems to be restricted to the representation of different kinds of variables with the corresponding constraints. Compared to the approach presented in this chapter, the problem of representing configuration knowledge on an abstract level is solved by providing a graphical representation of problem variables rather than using representations which are decoupled from a specific problem representation. An overview of different representation formalisms used in knowledge-based configuration is given in Stumptner (1997).

In Tiihonen, *et al.* (1999) a set of modeling concepts for building configuration models on a conceptual level is presented which is similar to the modeling concepts presented in Soininen, *et al.* (1998). In addition to these modeling concepts a set of modeling guidelines is proposed in order to support a correct application of the concepts. Compared to our approach, no concepts for representing constraints are defined, furthermore no clear semantics for the modeling concepts are given.

In the recent years several fields in AI focused research on the improvement of inter-operability of knowledge-based systems by developing ontologies. In Chandrasekaran, *et al.* (1999) an ontology is defined as a theory about the sorts of objects, properties of objects, and relationships between objects that are possible in a specified domain of knowledge. An earlier definition was given by Neches, *et al.* (1991) who define an ontology as follows: "An ontology defines the basic terms and relations comprising the vocabulary of a topic area as well as the rules for combining terms and relations to define extensions to the vocabulary". Also a well known definition of the term ontology is given in Gruber (1993) who defines an ontology as an "explicit specification of a conceptualization". Cranefield,Purvis (Cranefield and M., 1999) propose the application of UML as ontology construction language as an alternative to the AI approaches such as KIF (Neches, *et al.*, 1991). The usage of UML is motivated by the graphical representation of the modeling concepts which is more intuitive than a linear textual representation. Furthermore the existence of a large group of users which is already familiar with the UML is seen as a strong argument for the application of the language. Compared to the approach presented in this chapter there is no support for automatic translation of ontologies into an executable representation. In Robbins, *et al.* (1998) an approach is presented for applying UML for modeling software architectures. The built-in extension mechanisms of UML are applied in order to define an ADL[9]-specific meta-model inside UML, i.e. a UML profile for exchanging architecture descriptions. In this context OCL is used to formulate constraints (well-formedness rules) on the meta-model.

In Bourdeau and Cheng (1995) a formal semantics for OMT (Rumbaugh, *et al.* 1991) object model diagrams is given. A formal description of the system architecture containing the allowed states of the system is generated, which can be used to support verification and validation tasks. The main focus of this work is the generation of a formal specification in order to support the assessment of requirement specifications, whereas our approach concentrates on the definition of translation rules which support the automatic generation of a knowledge base which can be directly incorporated into a configuration environment.

The requirement for concepts supporting a precise definition of constraints on object-oriented models motivated the development of the Object Constraint Language (OCL (Warmer and Kleppe, 1999)). However, the semantics of the language is semiformal, i.e. it is defined in terms a context free grammar and a set of additional textual descriptions (Rational Software, *et al.*, 1997). In order to support e.g. testing a specification (represented in OCL) against an actual implementation, a precise semantics of the specification language is a prerequisite. In Richters and Gogolla (1998) a precise definition of the OCL is presented which similar to an algebraic specification. Baar, *et al.* (2000) discuss experiences in the application of OCL in industrial software development processes. In principle, OCL seems to be quite useful and software engineers and even domain experts with a technical background are able to apply OCL for stating formal constraints on a given object-model. Especially software engineers accepted OCL because of the similarities of its syntax to object-oriented programming languages. However, Baar, *et al.* (2000) point out that additional, more intuitive concepts are needed in order to support an effective introduction of OCL constraints. They made the observation that software engineers tried to change an objects state, what is prohibited by the declarative semantics of OCL. In order to tackle this challenge, Baar, *et al.* (2000) introduce the notion of constraint schemes. These schemes represent parameterizable constraints, which can be differently instantiated depending on the actual situation. For example, a constraint schema could restrict the occurrence of objects of a class to an upper bound. In this case the upper bound is represented by a variable which must be instantiated in order to instantiate the corresponding OCL constraint.

Mechanisms for structuring knowledge bases in the configuration domain are discussed in Fleischanderl and Haselböck (1993), where concepts similar to packages are applied in order to structure large configuration knowledge bases. Mechanisms for structuring configuration knowledge bases are also discussed in Feldkamp, *et al.* (1998), where the SyDeR approach is presented. For resolving the problem of defining complex constraints over the product structure, the concept of an interface is introduced. Using

---

[9] Architecture Description Language.

interfaces, complex constraints are not assigned to components, but are part of the interface which describes a connection between several components. Compared to this approach of using explicit interface definitions in order to organize complex constraints, contextual diagrams reorganize constraints themselves by using preconditions as structuring criteria.

7.7 CONCLUSIONS AND FUTURE WORK

In this chapter we have presented the basic concepts for improving the configuration knowledge base development process by applying and extending the basic modeling concepts provided by the Unified Modeling Language (UML). The approach enhances the application of Software Engineering techniques to knowledge-based systems by providing a knowledge-acquisition front-end for knowledge-based configuration systems. Extensible standard design languages like UML are able to provide a basis for introducing and applying rigorous formal descriptions of application domains. These languages are comprehensible and are widely adopted in established industrial software development processes. The automated generation of specialized software applications allows rapid generation of prototypes and furthermore improves the requirements engineering phase through short feedback cycles. Software development departments are enabled to incorporate formal description languages into their standard development process. The application of systems based on such languages is no more restricted to specialists with corresponding knowledge in the area of formal description languages. The design model is comprehensible for domain experts as well and can be adapted and validated without the need of specialists. By applying domain-specific modeling concepts domain experts are enabled to acquire configuration knowledge without having knowledge in modeling configuration knowledge bases using proprietary representations of configurator languages. This significantly contributes to a reduction of the knowledge acquisition bottleneck. Since the configuration knowledge base is represented on a conceptual level, configuration knowledge bases maintenance can be realized as well on a conceptual level, which leads to an increased degree of maintainability.

The application of the graphical modeling concepts of UML has its limits when building configuration knowledge bases[10], since in most domains there exist constraints which can not be represented graphically. The Object Constraint Language (OCL) is a language integrated in UML which was exactly developed for this purpose. In this work we have shown how to apply OCL for the formulation of complex constraints on UML configuration models.

The increasing size and complexity of configuration knowledge bases requires the provision of adequate modeling and structuring mechanisms. When modeling highly variant products which offer the customer a vast amount of interdependent options and choices, expressiveness restrictions are approached when using one single diagram. When acquiring configuration knowledge one has to cope with an intermingled and interdependent structural and functional product architecture. Exactly this challenge makes additional graphical representation concepts necessary. Motivated by the high complexity of large configuration knowledge bases we proposed a structuring mechanism called contextual diagrams.

There are a couple of further issues in the context of developing knowledge-based configuration systems, which we consider as a subject of future work. Knowledge acquisition using a conceptual modeling language (such as the UML) has shown to

[10] In other domains as well.

increase the applicability of knowledge-based configuration systems. However, the question has to be answered how to extend this approach in order to integrate additional and alternative concepts, such as natural language interfaces, which would allow the domain expert to explain the product structure and corresponding constraints in a more intuitive way without having knowledge in applying conceptual modeling techniques. The development of configurator user interfaces is strongly correlated to the development of configurator knowledge bases, since changes in the knowledge base trigger change requests for the corresponding user interface. Consequently, the support of the automatic construction of the configurator user interface implies corresponding reductions in development efforts. As has been shown in this chapter, modeling techniques from the area of Software Engineering have great potential in increasing the applicability of formal methods. Similar to interface generators in 4GL development environments, configurator user interfaces could be generated from the conceptual representation of the configuration knowledge base, i.e. an UML model of a configurable product can be further used to automatically generate the corresponding configurator user interface. The provision of software supporting the automatic generation of configurator user interfaces must be seen as a first step towards the realization of a sophisticated presentation layer for complex products and services. Different types of users (inexperienced, expert, etc.), different interaction styles (configuration from scratch, case-based configuration), and different preferences concerning products (customer is interested in multimedia personal computers, customer is primarily interested in server systems) impose the requirement for developing concepts for adaptive, i.e. personalisable user interfaces for configuration systems. The question has to be answered, whether the presented approach to knowledge acquisition can be enhanced in order to consider personalization concepts as well. The vision in this field could be to build a "virtual salesman" (some kind of configuration agent), which actively supports a personalized configuration process, i.e. determines the way the product is presented to the customer and the way the configuration system interacts with the customer.

One major task in the context of distributed configuration is the identification of concepts for integrating different configuration systems based on different knowledge representation formalisms. The component-port representation used in our work for formalizing configuration models built in UML has shown to be a widely accepted representation formalism for configuration problems. However, a general architecture for integrating different configurators has to be identified, which supports the integration of heterogenous configuration systems as well. Agent-based approaches for integrating knowledge-based systems seem to be a promising approach for the configuration domain. Agent communication languages (such as KQML (Finin, 1997)) support the communication between heterogenous knowledge-based systems, furthermore the corresponding knowledge representation languages enable the construction of ontologies, which support knowledge interchange between the engaged systems. The challenge here is to define ontologies suitable for integrating a large set of configuration systems.

The Semantic Web (Berners-Lee, 1999) is the vision of developing enabling technologies for the Web which support access to its resources not only to humans but as well to applications often denoted as agent-based systems providing services such as information brokering, information filtering, intelligent search or synthesis of services. The technology provided by the Semantic Web community seems to be applicable for the configuration domain as well – especially for describing the capability of configuring certain product types as a kind of Web service and as a basis for defining common ontologies used as communication basis in distributed configuration.

BIBLIOGRAPHY

Akman, V., and Surav, M., Steps Towards Formalizing Context: In *AI Magazine*, Vol. 17, pp. 55-72 (1996).

Aldanondo, M., Moynard, G. and Hamou, K.H.: General configurator requirements and modeling elements, In *Workshop on Configuration*, edited by Stumptner, M., (Berlin), pp. 1-6 (2000).

Anderson, D.M.: Agile Product Development for Mass Customization, McGraw-Hill (1997).

Baar, T., Hähnle, R., Sattler, T. and Schmitt, T.H.: Entwurfsmustergesteuerte Erzeugung von OCL Constraints, In *Informatik* 2000, edited by Mehrhorn, K. and Snelting, G., Vol. 30, Jahrestagung der Gesellschaft für Informatik, (Springer), pp. 389-404 (2000).

Barker, V.E., O'Connor, D.E., Bachant, J.D. and Soloway, E.: Expert systems for configuration at Digital: XCON and beyond, In *Communications of the ACM*, Vol. 32, No. 3, pp. 298-318 (1989).

Berners-Lee, T.: *Weaving the Web*, (Orion Business Books) (1999).

Booch, G.: Object-Oriented Analysis and Design with Applications, In *Addison Wesley Object Technology Series* (1994).

Bourdeau, R.H. and Cheng, B.: A Formal Semantics of Object Models, In *IEEE Transactions on Software Engineering*, Vol. 21, No. 10, pp. 799-821(1995).

Cameron, J.:JSP&JSD: *The Jackson Approach to Software Methodology, 2nd edition*, IEEE Computer Society (1989).

Chandrasekaran, B., Josephson, J. and Benjamins, R.: What Are Ontologies, and Why do we Need Them? In *IEEE Intelligent Systems*, Vol. 14, No. 1, pp. 20-26 (1999).

Charlton, C. and Wallace, K:Reminding and context in design, In *Proceedings 6th International Conference on Artificial Intelligence in Design* (AID'oo), (Boston: Kluwer Academic Publishers), pp. 569-588 (2000).

Cook, S. and Daniels, J.:Designing Object Systems – Object Oriented Modeling with Syntropy, (Prentice Hall) (1994).

Cranefield, S. and M. S. Purvis: UML as an Ontology Modelling Language, In *Proceedings of the Workshop on Intelligent Information Integration*, (16th International Conference on Artificial Intelligence, Stockholm) (1999).

Delisle, N.M. and Schwartz, M.D.: Contexts – a partitioning concept for hypertext, In *ACM Transactions on Information Systems*, Vol. 5, No. 2, pp. 168-186 (1987).

Diller, A.: Z – An Introduction to Formal Methods, John Wiley & Sons (1994).

Feldkamp, F., Heinrich, M. And Gramann, M.: SyDeR System Development For Reusability, In *AIEDAM* Vol. 12, No. 4, pp. 373-382 (1998).

Felfernig A. and Zanker, M.: Diagrammatic Acquisition of Functional Knowledge for Product Configuration Systems with the Unified Modeling Language, In *Proceedings to International Conference on Theory and Application of Diagrams* (Diagrams'2000), Springer Lecture Notes in Artificial Intelligence, Vol. 1889, (Edinburgh), pp. 361-375 (2000).

Felfernig, A., Friedrich, G. and Jannach, D.: UML as domain-specific language for the construction of knowledge-based configuration systems. International Journal of Software Engineering and Knowledge Engineering (IJSEKE), 10,4: pages 449-469 (2000-a).

A. Felfernig, G. Friedrich, and D. Jannach: Generating product configuration knowledge bases from precise domain extended UML models, In *Proceedings 12th International Conference on Software Engineering and Knowledge Engineering*, (Chicago), pp. 284-293 (2000-b).

Felfernig, A., Friedrich, G., Jannach, D. and Stumptner, D.: Consistency-Based diagnosis of Configuration Knowledge Bases, In *Proceedings of the 14th European Conference on Artificial Intelligence* (ECAI'2000), (Berlin), pp. 146-150 (2000-c).

Felfernig, A., Jannach, D. and Zanker, M.: Contextual Diagrams as Structuring Mechanisms for Designing Configuration Knowledge Bases in UML. In *Proceedings of 3rd International Conference on the Unified Modeling Language* (UML'2000), Springer Lecture Notes in Computer Science, Vol. 1939, (York), pp. 240-254 (2000-d).

Finin, T., Labrou, Y. and Mayfield, J.: KQML as an agent communication language, In *Software Agents*, edited by Bradshaw, J., (Cambridge: MIT Press) (1997).

Fleischanderl, G. And Friedrich, G.: A. Haselböck, H. Schreiner, and M. Stumptner, Configuring Large Systems Using Generative Constraint Satisfaction, In *IEEE Intelligent Systems, Special Issue on Configuration*, Vol. 13, No. 4, edited by Freuder, E. and Falting, B., (IEEE) pp. 59-68 (1998).

Fleischanderl, G. and Haselböck, A.: Thoughts on Partitioning Large-Scale Configuration Problems, In *AAAI Fall Symposium on Configuration*, edited by Faltings, B. and Freuder, E., pp. 45-54 (1996).

Friedrich, G. And Stumptner, M.: Consistency-Based Configuration, In *AAAI Workshop on Configuration, Technical Report* WS-99-05, (Orlando), pp. 35-40 (1999).

Gruber. T.: A translation approach to portable ontology specifications, *Knowledge Acquisition*, pp. 199-220 (1993).

Jacobson, I., Christerson, M. and Övergaard, G.: Object-oriented Software Engineering – A Use-Case Driven Approach, (Addison Wesley) (1992).

McCarthy, J., 1993: Notes on formalizing context. In *Proceedings of the 13th IJCAI*, (Chambery, France) pp. 555-560 (1993).

Mittal, S. and Frayman, F.: Towards a Generic Model of Configuration Tasks,. In *Proceedings of the 11th IJCAI*, (Detroit) pp. 1395-1401 (1989).

Mittal, S. and Falkenhainer, B.: Dynamic Constraint Satisfaction Problems, In *Proceedings AAAI 1990*, (Boston), pp. 25-32 (1990).

Neches, R., Fikes, R., Finin, T., Gruber, T., Patil, Senator, T. and Swartout, W.: Enabling technology for knowledge sharing, In *AI Magazine*, Vol. 12 No. 3, pp. 36-56 (1991).

Rational Software, Microsoft, Hewlett Packard, Oracle, Sterling Software, MCI Systemhouse, Unisys, ICON Computing, Intellicorp, I-Logix, IBM, ObjecTime, Platinum Technology, Ptech, Taskon, Reich Technologies, and Softeam, OCL Specification ver. 1.1. (1997).

Richters, M. and Gogolla, M.: On Formalizing the UML Object Constraint Language OCL, In *Proceedings 17th International Conference on Conceptual Modeling* (ER'98) (1998).

Rumbaugh, J., Blaha, M., Premerlani, W., Eddy, F. and Lorensen, W.: Object-Oriented Modeling and Design, In *Prentice Hall International Editions*, (New Jersey) (1991).

Rumbaugh, J., Jacobson, I. and Booch, G.: The Unified Modeling Language Reference Manual, (Addison Wesley) (1998).

Robbins, J.E., Medvidovic N., Redmiles, D.F. and Rosenblum D.S.: Integrating Architecture Description Languages with a Standard Design Method, In *Proceedings 20th ICSE*, (Kyoto) pp. 209-218 (1998).

Schreiber, A.T., Wielinga, B.J., DeHoog, R., Akkermans, H. and van de Velde, W.: CommonKADS: A Comprehensive Methodology for KBS Development, In *IEEE Transactions on Knowledge and Data Engineering*, Vol. 9, No. 6, pp.28-37 (1994).

Siegel, M., Sciore, E. and Salveter, S.: A Method for automatic rule derivation to support semantic query optimization, In *ACM Transactions on Database Systems*, Vol. 17, pp. 563-600 (1992).

Soininen, T., Tiihonen, J., Männistö, T. and Sulonen, R.: Towards a General Ontology of Configuration, in AIEDAM, Vol. 12, No. 4, pp. 357-372 (1998).

Stumptner. M.: An overview of knowledge-based configuration, In *AI Communications*, Vol. 10, No. 2 (1997).

Stumptner, M. and Haselböck. A.: A generative constraint formalism for configuration problems, In *Advances in Artificial Intelligence: Proceedings of the 3rd Congress of the Italian Association for Artificial Intelligence* AI*IA'93, edited by Torasso, P., (Springer, Berlin, Heidelberg) pp. 302-313 (1993).

Theodorakis, M., Analyti, A., Constantopoulos, P. and Spyratos, N.: Context in information bases, In *Proceedings of 3rd International Conference on Cooperative Information Systems* (CoopIS'98), (New York City: IEEE Computer Society), pp. 260-270 (1998).

Tiihonen, J., Lehtonen, T., Soininen, T., Pulkinen, A., Sulonen, R. and Riitahuhta, A.: Modeling Configurable Product Families, In *Proceedings of the 12th International Conference on Engineering Design* (ICED'99), edited by Lindemann, U., Birkhofer, H., Meerkamm, H. and Vajna, S., (München), pp. 1139-1142 (1999).

Unified Modeling Language Specification Ver. 1.3 June (1999).

Warmer, J. and Kleppe, A.: The Object Constraint Language – Precise Modeling with UML, In *Addison Wesley Object Technology Series* (1999).

CHAPTER 8

A Domain Specific Repository and Reuse for Computer-Assisted Instruction Systems

Takashi Fuji

Hokkaido Information University

Kiyoshi Itoh

Sophia University

ABSTRACT

This chapter describes a domain specific repository based on the methodology for Computer-Assisted Instruction (CAI) systems development. The methodology, which is based on information engineering and object-oriented technology, is used to support navigation for courseware development, designing the curriculum of CAI components, and CAI systems resources management. A repository-based approach is key to support of CAI systems development and reuse of CAI systems resources such as application frameworks and components. By the structure of the repository such as meta-model / model / object base and by use of object-oriented modeling, it is constructed for reuse by systems analysts and subject matter experts. Systems analysts can build new models by reusing meta-model and models for their target system. Subject matter experts can develop their courseware by reusing models and object bases for their CAI systems.

Since the Internet is used worldwide, learning environment is also changing to incorporate use of the Internet. In various types of organizations, such as universities and private enterprises, subject matter experts are developing various types of courseware; each site can be connected to the network and used to make a variety of courseware. By using CORBA architecture, we developed the distributed CAI systems development environment.

8.1 INTRODUCTION

Software reuse, a subject of long-standing interest, has currently emerged in a new, business-critical guise. Software reuse is effective to achieve the goal of a software organization, which is to build useful working software systems in a timely manner and within given economic constraints. In particular, application frameworks and components based on object-oriented technology are the main reusable elements in reuse projects such as the REBOOT ("REuse Based on Object-Oriented Techniques") project (Karlsson, 1995) in the ESPRIT III.

In the domain of educational systems such as CAI systems, instructional design theory (Merril, 1987), (Gagné, 1993) is being used as a method of developing instructional material. Systems such as ID Expert (Merril, 1987), IDE (Russel *et al.*, 1988), and GAIDA (Gagné, 1993), have been developed by instructional design automation. These systems provide a set of tools to be used in the instructional process. However, these systems do not have a mechanism for reuse of CAI systems resources such as application frameworks and pedagogy components and for collaboration working. These systems are constructed mainly for reuse of instructional units and for curriculum conception.

The purpose of this chapter is to present the methodology and development environment of CAI systems for the systems analyst and the subject matter expert according to the domain specific repository with object-oriented technology. For the systems analyst, REBECCA (Repository-based Environment for Computerized Courseware Analysis) (Fuji 1997) enables the reconstruction of CAI systems by reusing application frameworks such as AVALON (Audio Visual and Adaptive Learning system by Oneself in Network) (Fuji *et al.*, 1994) and CAMELOT (Collaborative and Multimedia Environment for Learners on Teams) (Fuji *et al.*, 1996). For the subject matter expert, WEB-CAT (WWW-based Curriculum Authoring with Thesaurus) (Fuji and Hoshihara, 1997) enables the construction of the intended curriculum with collaborative work through WWW, while REBECCA provides a means of course navigation, the reuse of instructional material, and target system generation.

As learning styles with CAI systems, there are individual learning environment and collaborative learning environment. We have developed prototypes of an application framework, AVALON, for the individual learning environment and CAMELOT for the collaborative learning environment. These application frameworks are constructed by use of object-oriented technology, and are stored in the domain specific repository called REBECCA for reuse. Repository-based methodology is derived from information engineering (Martin, 1989). With regard to the components in CAI systems, there are instructional material, curriculum, and pedagogy. We have developed a prototype called WEB-CAT for developing curriculum through collaborative work with a domain specific repository and WWW. And, as the REBECCA is constructed by an object-oriented technology, it is easy to adapt CORBA (Common Object Request Broker). CAI components and CAI frameworks of the distributed REBECCA correspond to a CORBA application object, and these components use CORBA services for searching distributed object reference by name or for creating new objects.

This chapter is organized into three sections: Section 8.2 outlines the framework in CAI systems; Section 8.3 introduces the basic concepts of the domain specific repository; Section 8.4 describes the distributed CAI systems development environment with CORBA architecture.

8.2 A FRAMEWORK AND COMPONENTS FOR CAI SYSTEMS

The reuse approach is based and relies on the concepts that reusable software components are produced and stored in a repository. There are two main types of object-oriented components:

1. solution-based inheritance hierarchies, and
2. domain-based frameworks 1). This framework is based on domain-based frameworks such as CAI systems, and we use a framework in the sense of a skeleton application that the developer can adjust with it. The term 'component' is

often used interchangeably. Software component is a physical packaging of executable software with a well-defined and published interface (Hopkins, 2000). The following definition of component will be used in this chapter: "core elements that compose an application framework."

8.2.1 Methodology for CAI Systems Development with Repository

In software engineering, structured techniques are applied to one project, whereas in information engineering, structured techniques are applied to the enterprise as a whole or to a large sector of the enterprise. Information engineering techniques encompass those of software engineering in a modified form. Information engineering has been defined with reference to automated techniques as follows: an interlocking set of automated techniques in which enterprise models, data models and process models are built up in a repository and are used to create and maintain data-processing systems. The repository is a key technology in information engineering, and it steadily accumulates information relating to the planning, analysis, design, construction and, later, the maintenance of systems.

The field of instructional design was initiated by the United States Department of Defense during World WarII to improve the efficiency of certain types of high level technical training programs such as pilot training (Reiser, 1987). The systems approach is a step-by-step procedure for solving problems. It has been applied to educational problems at all levels, from education in general, to training for a specific target population. In instructional design, the systems approach is sometimes referred to as instructional systems development. There are numerous models demonstrating this approach. Although the models differ in detail, such as the number and sequence of steps in which specific components are grouped, the common basic steps are grouped into the following five categories: analysis, design, production, implementation, and evaluation.

The methodology of REBECCA is based on that of information engineering and instructional design. The characteristics of this methodology are as follows.

- Object-oriented repository for managing CAI systems resources
- Reuse for CAI systems framework (individual learning / collaborative learning)
- Reuse for CAI systems components (instructional material, curriculum, pedagogy)
- Navigation for courseware development (analysis, design, construction)

8.2.2 CAI Frameworks

The CAI systems frameworks used are: AVALON for the individual learning environment and CAMELOT for the collaborative learning.

AVALON

The AVALON CAI framework is a multimedia intelligent tutoring system. The system provides adaptive instructional units and pedagogy suited to the individual learner's understanding. The system also provides multimedia instructional units, such as video, voice, text, and image data, suited to each individual learner's characteristics. This framework, composed of pedagogy, a learner model and instructional material information, was used as a base (Figure 8.1). In order to realize the following two functions, i.e., a function for dynamic representation of multimedia instructional material corresponding to the comprehension level of each learner and to the media characteristics, and a function for interactive response to questions from learners, we developed an instructional object in which the multimedia instructional material and knowledge expression of instructional material were encapsulated as an instructional object.

As shown in Figure 8.1, AVALON is composed of "Learner" (class) and "Intelligent Tutor" (class). "Learner" has "Characteristic", "Learning Goal", and "Student Model". A part of "Intelligent Tutor" is composed four tiers such as *strategy, task, reasoning*, and *domain*. The *strategy* tier is "Intelligent Tutor", and it manages learning process. There are "Agent for Learners" for facilitating learner's motivation and "Pedagogy" for deciding how to teach in *task* tier. "Pedagogy" is composed of "Strategy" and "Tactics", and these are in the *reasoning* tier. "Strategy" decides learning process by selecting "Curriculum" and "Tactics", and presents adaptive instructional material by selecting "Instructional Objects". The *domain* tier is composed of "Curriculum" and "Instructional Objects".

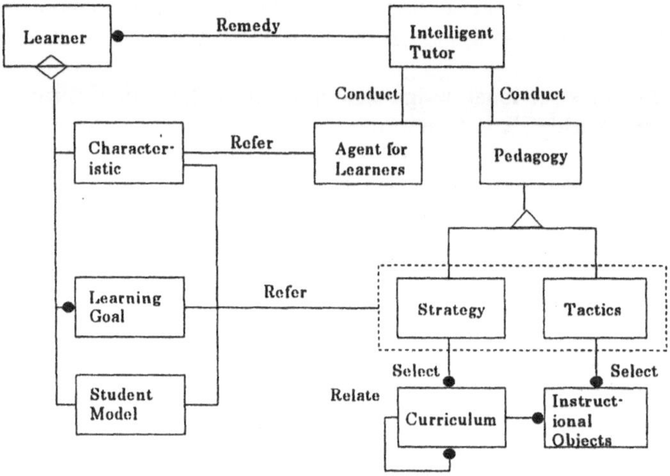

Figure 8.1 Object Model of CAI Framework for Individual Learning

CAMELOT

The CAMELOT CAI framework is a collaborative learning support system. The learning process is based on a nominal group technique. The system facilitates collaborative learning navigation, group discussion, and remedy of collaborative work solutions. The pedagogy is a combination of rule-based reasoning and case-based reasoning. In the business systems design environment, there may be more than one solution to any given problem. For example the data model will be different depending on the perspective of each individual leaner.

Thus, collaborative learning systems are very effective in this domain. By using CAMELOT, each learner learns how to analyze through case study and how to cooperate with his or her group in problem solving. Learners come to a deeper understanding from using CAMELOT than from studying independently, because of shared discussion, tips from other learners, and examinations, and review of one another's works. Figure 8.2 shows a functional model of CAI framework for a collaborative learning environment. Figure 8.3 shows a sample screen.

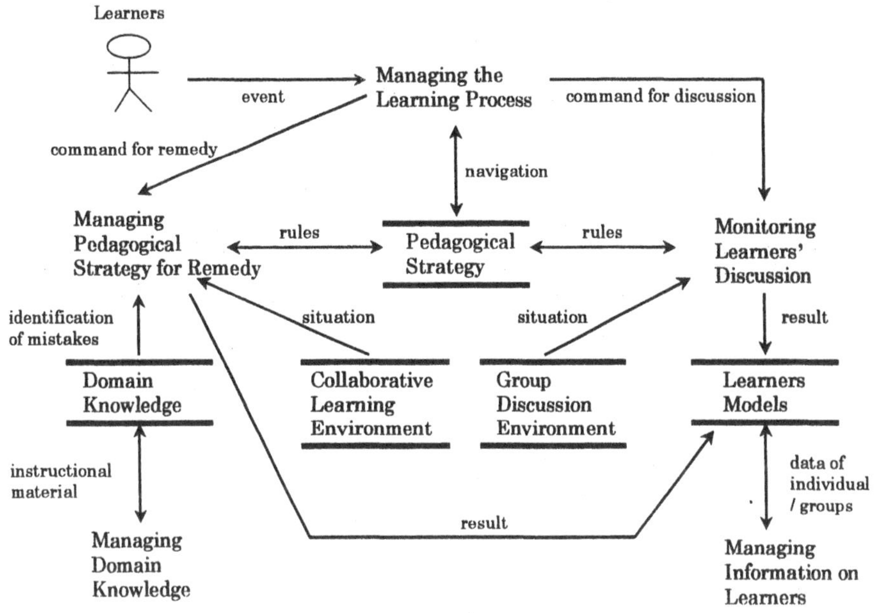

Figure 8.2 Functional Model of CAI Framework for Collaborative Learning

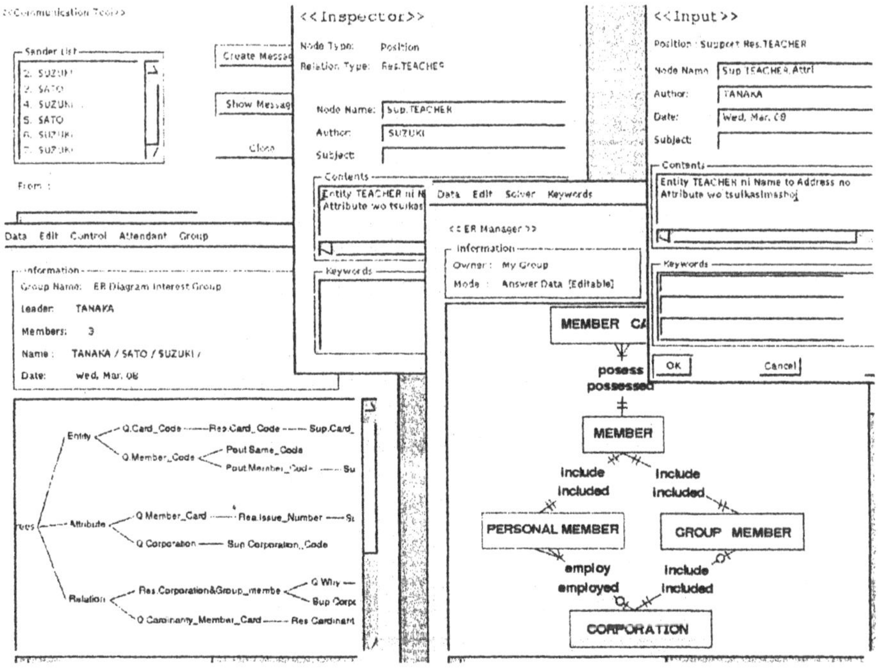

Figure 8.3 Sample screen for Collaborative Learning Environment (CAMELOT)

Domain knowledge is expressed with a hierarchical model, which is divided into multiple domain worlds in the form of the tree. The root of the tree is a world called the domain universe, and other worlds correspond to certain aspects of the knowledge of the domain universe according to the dependency relationship of the knowledge, logical chunking of the knowledge, the possible scope of activity that learners can derive from the worlds, and so on.

Learner models are composed of the individual learner model and the group learner model. The individual learner model includes two kinds of information:

(1) the type and content of each individual's mistakes, and
(2) no participation, occupation by an individual learner, and type of activity (e.g., issue, answer, agree, disagree).

The group learner model also has two kinds of information:

(1) the type and content of mistakes for group work, and
(2) the duration and the number of times of no participation.

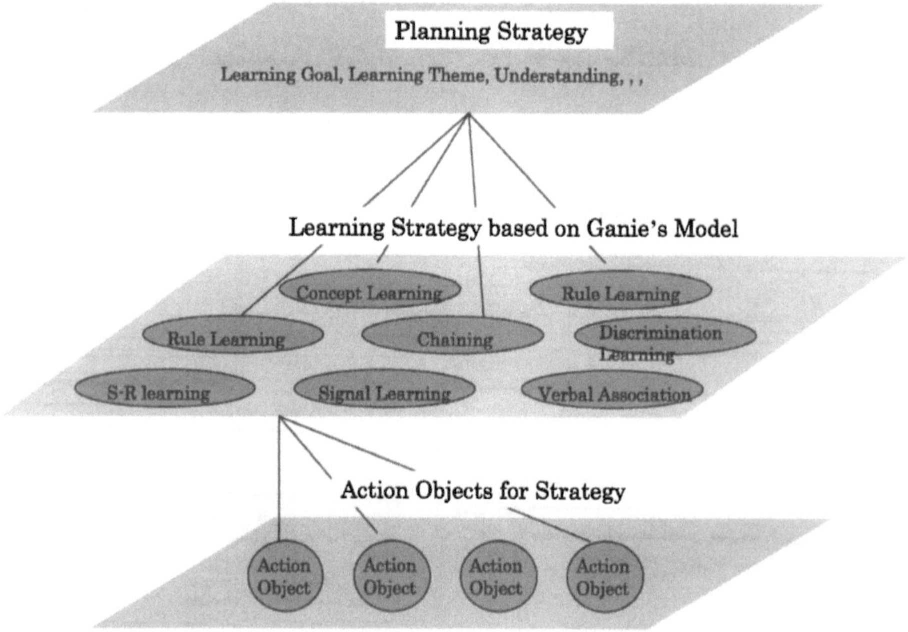

Figure 8.5 Structure of Pedagogy Object

8.2.3 CAI Components

As domain specific components in CAI systems, there are the instructional object, curriculum object, and pedagogy object that compose CAI application framework.

Instructional Object

An instructional object is an extended instructional unit. An instructional object is composed of a multimedia object and a knowledge object. A multimedia object incorporates multimedia instructional material, such as text, image, voice, and video. A knowledge object incorporates the knowledge representation of instructional units, such as semantic network, rule-based reasoning, and case-based reasoning (Figure 8.4). An instructional object is used for components when subject matter experts design courseware.

Curriculum Object

A curriculum object is related to instructional objects and selected by pedagogy objects. A curriculum object is structured using a thesaurus. This object is composed of a superordinate concept, a subordinate concept, and related concepts.

Pedagogy Object

A pedagogy object is composed of three tiers such as planning strategy, learning strategy, and action objects for strategy (Figure 8.5). The basic concept of the learning strategy is based on Gagné's accumulated learning model (Gagné, 1985). The learning strategy is composed of "problem solving", "rule learning", "concept learning", "multiple discrimination learning", "verbal association", "chaining", "stimulus-response learning", and "signal learning". Subject matter experts can design their pedagogy for CAI systems by combining these pedagogy objects .

8.3 REPOSITORY FOR CAI SYSTEMS DEVELOPMENT

CAI systems resources are stored in the repository and reused through the repository. One of the major function of the repository is managing CAI systems resources such as conceptual modeling, consistency management, and browsing repository. There is another major function, supporting CAI systems development, such as navigation for courseware development, designing the course, and designing the curriculum with collaboration.

8.3.1 The Function of Repository

Conceptual modeling

Conceptual modeling is used by systems development people such as the systems analyst. When a CAI system is to be developed, CAI application frameworks, such as AVALON and CAMELOT, are used to create a skeleton of a CAI system. Systems analysts modify the skeleton by using application frameworks, and REBECCA supports a construction of a conceptual model using the object modeling technique (OMT) (Rumbaugh, 1991). As a meta-model for modeling a conceptual model, OMT is used.

Consistency management

CAI systems resources in the repository are necessary to maintain consistency for the object. If systems analysts modify the frameworks using symbols, which are not defined in the meta-model, the system checks the consistency between meta-model and the frameworks.

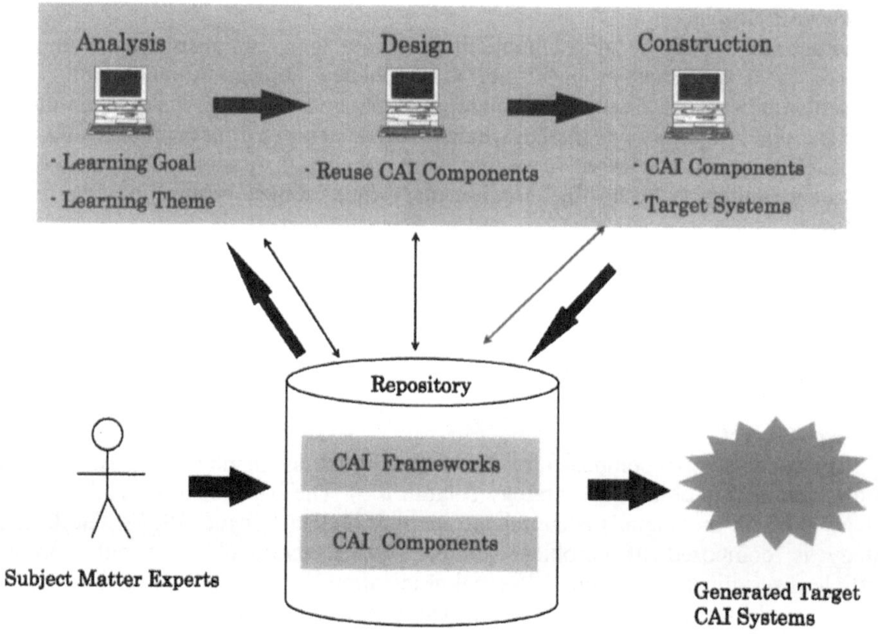

Figure 8.6 Process Navigation for Courseware Development

Browsing repository

By means of a browsing function, the user can view the contents of the repository. Systems analysts can view the structure of the frameworks, and subject matter experts can view the reusable components such as instructional material, curriculum, and pedagogy.

Navigation for courseware development

The steps of CAI systems development are analysis, design, and construction. Subject matter experts define the learning department, and specify a learning goal and theme in the analysis phase. In the design phase, they specify reusable CAI resources, such as pedagogy object, curriculum object, and instructional object. They can locate the reusable resources by using the thesaurus. In the construction phase, they produce the instructional object that cannot be reused, and finally, the system generates the target CAI system required by the subject matter expert. The system navigates these procedures based on the methodology for CAI systems development with repository. Figure 8.6 shows the process flow of CAI systems development for the subject matter expert.

Designing the course

The subject matter expert can reuse the curriculum stored in the repository, and add new themes or delete unnecessary themes. At the same time, he or she can retrieve instructional objects related to the theme by using the thesaurus. This thesaurus has been developed for information systems education in universities.

Designing the curriculum with collaboration through the WWW

In defining the curriculum, especially in new domains such as information systems education, collaboration is useful. REBECCA supports collaborative work on the Internet using a discussion model (Fuji and Tanigawa, 1997) in order to define a new curriculum. Using the WWW, subject matter experts from various countries can define a curriculum while sharing ideas. The defined curriculum is stored in the repository.

8.3.2 The Structure of the Repository

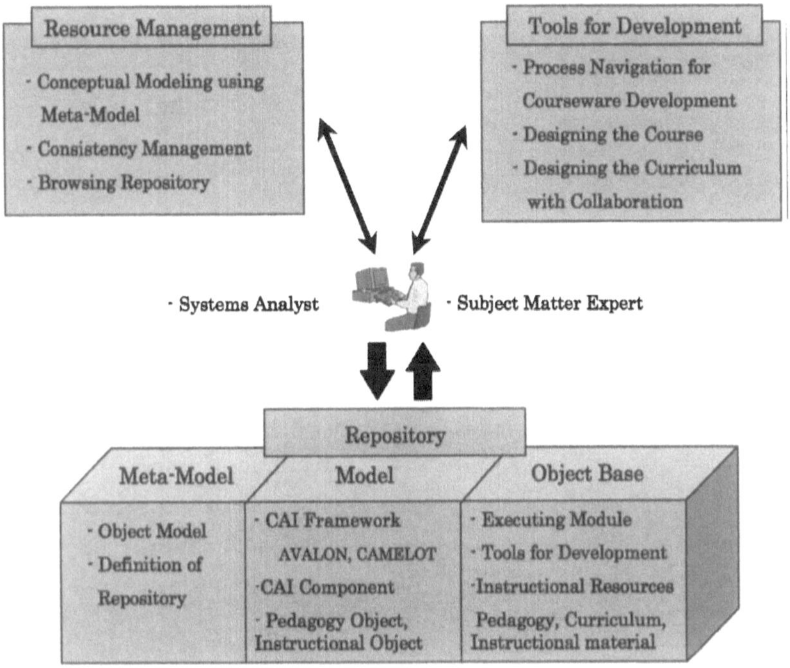

Figure 8.7 Structure and functionality of the repository

A repository is a place for storing all the information required in a software engineering environment; for example: tools, software products and documents. The Portable Common Tool Environment (PCTE) is a standard for a public tool interface for an open repository (Wakeman and Jowerr, 1993). However a meta-model of the PCTE is defined by extended Entity-Relationship modeling and is not defined encapsulation. The meta-model of REBECCA is defined with an object model.

The Case Data Interchange Format (CDIF) has been developed to define the structure and content of a transfer that may be used to exchange data between two CASE tools. CDIF uses models to define the information content of transfer. Several layers of model are used, the second being an abstraction of the first, the third of the second, and so on. There are many different types of information that a user could want to transfer between CASE tools. Often, these types of information are identified by describing their location in a model of Systems Development Life Cycle (SDLC). There are many models of the SDLC. CDIF can support transfer of information that lies anywhere within the SDLC. The goal of the CDIF integrated meta-model is to provide definition of all information of interest to CASE tools.

The structure of a repository is based on the CDIF architecture. The repository of REBECCA is composed of three levels. Figure 8.7 shows the structure and functionality of the repository.

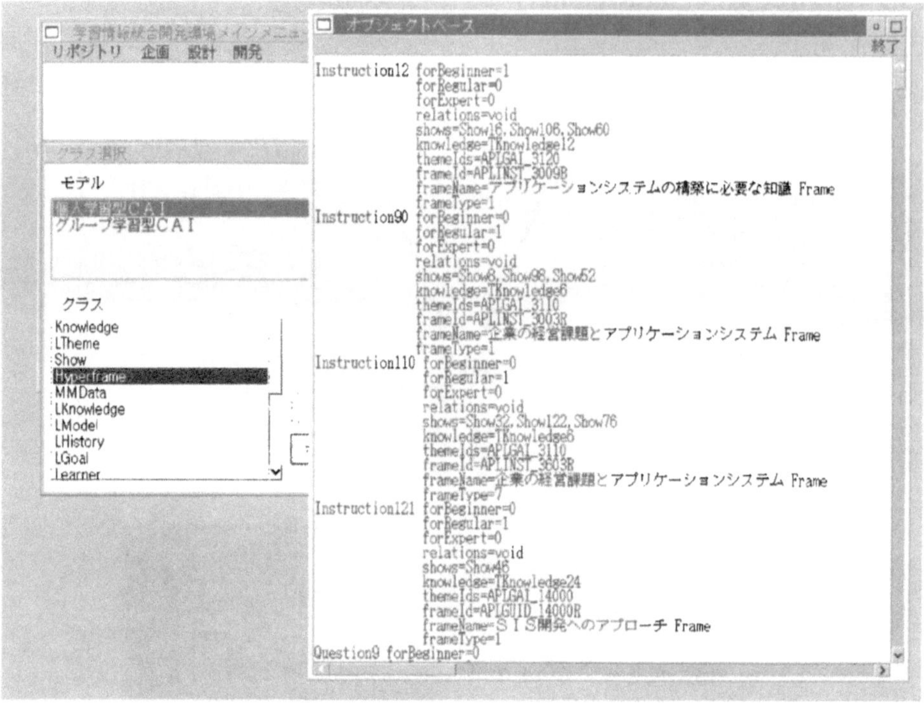

Figure 8.8 Sample Screen of Object Base in the Repository

Level 1: Meta-model

The first level holds a model of about level 2, i.e., a model about the model describing the resources at level 3. Models about models are commonly called meta-models. The semantic content of the meta-model is defined using notations of object modeling technique, such as class, association, generalization, and aggregation. This meta-model also self-describes the repository.

Level 2: Model

The second level in the repository is the conceptual model defined using the meta-model. Frameworks and components are described in the model. AVALON, the framework for individual learning environment, functions as a multimedia intelligent tutoring system. Figure 8.1 is a part of the model of AVALON. CAMELOT is the framework for the collaborative learning environment, and is described as a part of the model in Figure 8.2. The instructional object and pedagogy object are components that can be reused, and described also using meta-model.

Level 3: Object Base
The third level in the repository is the implemented object based on the level 2 model. There are executable CAI systems, development tools, and instructional resources in the third level of the repository (Figure 8.8).

8.4 DISTRIBUTED ENVIRONMENT WITH REPOSITORY

Since the Internet is used worldwide, the global learning environment also is changing to incorporate use of the Internet. In many types of organizations, such as universities and private enterprises, subject matter experts are developing various types of courseware; each site can be connected to the network and they can find a variety of courseware. However, they are not able to use courseware for reuse, because the various types of courseware are developed and executable from different platforms such as Pentium, PowerPC, SPARC, and so on. For the common use of the courseware through the Internet, the distributed heterogeneous development environment is required.

8.4.1 Utilization of Distributed Object Technology

Common Object Request Broker (CORBA) is an industry-wide standard for creating distributed object systems. It is a standard that is accessible from many different languages and allows interoperability on various platforms. Using CORBA, a client object can transparently invoke a method on a server object, which can be on the same machine or across a network. CORBA intercepts the call and is responsible for finding an object that can implement the request, pass on to it the parameters, invoke its method, and return the results. The client does not have to be aware of where the object is located, its programming language, or its operating system.

From these characteristics of CORBA, we developed a distributed development environment REBECCA using CORBA. CORBA provides the following benefits in development of distributed development environment:

- CORBA supports many programming language mapping
- CORBA supports distributed computing and object orientation
- CORBA provides interoperability between different vendor's ORBs

8.4.2 Distributed Repository Using CORBA

As the REBECCA is constructed by an object-oriented technology, it is easy to adapt CORBA. Object Management Architecture (OMA) has four major components, ORB (Object Request Broker), CORBA services, CORBA facilities, and application objects. ORB is a communication infrastructure between distributed objects. CORBA services are a set of system service interface, and include the naming service and the persistent service.

CORBA facilities represent specifications with great detail for the construction of the distributed system. CAI components of the distributed REBECCA correspond to a CORBA application object, and these components use CORBA services for searching distributed object reference by name or for creating new objects. CAI frameworks (such as AVALON and CAMELOT) correspond to CORBA facilities that provide domain dependent services. At each site, there is a repository object stored CAI component such as instructional objects, curriculum objects, and pedagogy objects. The repository object manages the other objects and communicates with the repository object of another site in order to obtain information from and access to resources at another site (Figure 8.9). The distributed REBECCA with CORBA provides the following functions:

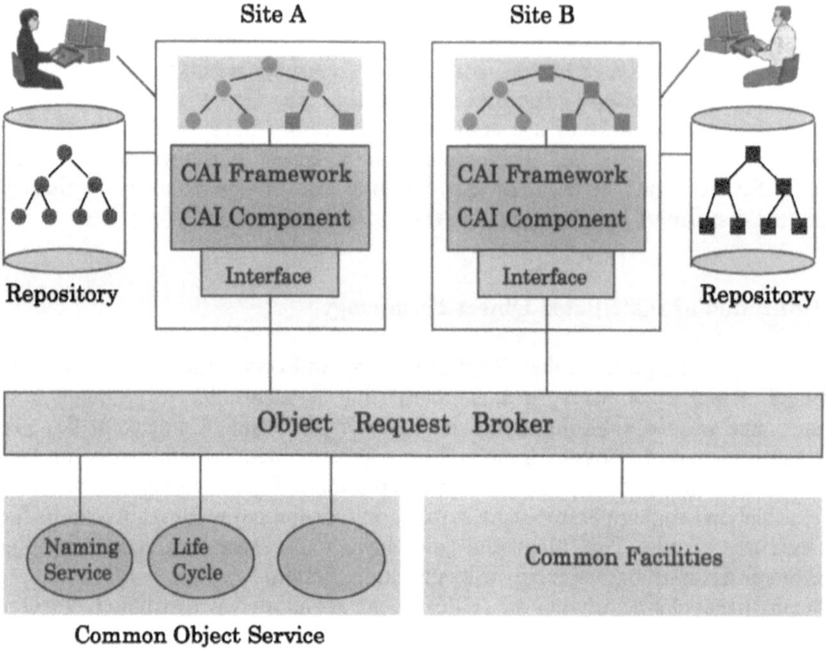

Figure 8.9 Distributed Development Environment with CORBA

Enables the reuse of heterogeneous and distributed CAI components

The distributed REBECCA provides methods to treat heterogeneous distributed CAI components uniformly, so that subject matter experts are able to reuse these CAI components in developing courseware. CAI systems can provide adaptive instructional objects on the network.

Appropriate site of CAI components

In using CORBA environment, implementation of CAI components is encapsulated and can be placed at the distribution site, according to the characteristics of each CAI components. Therefore, CAI components can be placed in the appropriate sites. For instance, an instructional object that includes the simulation material, which requires a large amount of computer power, is placed at a high performance computer site, and the pedagogy object that requires high inference performance is placed at a computer site with high performance engine for inference.

Management for distributed CAI components

The distributed repository manages each CAI components stored in each server site. The distributed CAI components are able to maintain consistency by means of global instruction manager (Tanigawa *et al.*, 1998).

8.5 CONCLUSION

In this chapter, we have described a methodology for CAI systems development and a domain specific repository for reuse of CAI framework and CAI component. The repository is the key to supporting CAI systems development and reusing CAI systems resources. Using a three level structure, such as meta-model, model, object-base, and using object-oriented modeling, REBECCA is constructed for ease of reuse for systems analysts and subject matter experts. Furthermore, we construct a distributed CAI systems development environment using CORBA architecture.

ACKNOWLEDGEMENT

This research is partly supported by the Japan Society for the Promotion of Science under the Grant-in-Aid for Scientific Research No. C-2-13680246. We thank Katherine W. Mansoor for her cooperation.

REFERENCES

Fuji, T., Tanigawa, T., Hoshihara, K., Fujii, K., Kozeni, M. and Saegusa, T., 1994, Development of Intelligent Multimedia CAI with Hyper-frame, In *Proceedings 1994 Informatics Symposium*, IPSJ, pp.133-141.

Fuji, T., Tanigawa, T., Kozeni, M., Inui, M. and Saegusa, T., 1996, A Case-Based Approach to Collaborative Learning for Systems Analyst Education, In *Proceedings ITS'96*, (Springer-Verlag), pp.177-186.

Fuji, T., 1997, A Repository-based Approach to Reuse Educational Systems Resources, In *Proceedings ED-MEDIA & ED-TELECOM97*, pp. 378-383.

Fuji, T., Tanigawa, T., Inui, H. and Saegusa, T., 1997, Using Case-Based Reasoning for Collaborative Learning System on the Internet, In *IEICE Transactions on Information and Systems*, Vol.80-D, No.2, pp.135-142.

Fuji, T., Hoshihara, K., Kozeni, M. and Saegusa, T., 1997, Developing a WWW-based Support System for Designing the Curriculum, In *IPSJ Journal*, Vol.38, No.11, pp.2370-2381.

Gagné, R.M., 1985 *The Conditions of Learning (4th edition)*, (New York: Holt, Rinehart, and Winston).

Gagné, R.M., 1993, Computer-based Instructional Guidance, pp.133-146, In *Automating Instructional design: Concept and Issues* edited by Spector, J.M., Polson, M.C., and Muraida, D.J., (Englewood Cliffs: Educational Technology).

Hopkins, J., 2000, Component primer, In *Communications of the ACM*, Vol.43, No.10, pp.27-30.

Karlsson, E.A., 1995, *Software Reuse*, (West Sussex: John Wiley & Sons).

Martin, J., 1989, *Information Engineering Book 1 Introduction*, (Prentice-Hall).

Merril, J., 1987, An expert system for instructional design, In *IEEE-Expert* Vol. 2 No. 2, pp.25-37.

Reiser, R. A., 1985, Instructional Technology: A history, pp.11-48, *Instructional Technology: Foundations* (Gagné, R. M.), (Hillsdale: Erlbaum Associates).

Rumbaugh, J., Blaha, M., Premerlani, W., Eddy, F. and Lorensen, W., 1991, *Object-Oriented Modeling and Design*, (Englewood Cliffs: Prentice Hall).

Russel, D., Moran T. and Jordan, D., 1988, The Instructional Design Environment, , In *Intelligent Tutoring Systems: Lesson Learned* edited by Psotka, J., Massey, L. and Mutter, S., (Hillsdale: Lawrence Erlbaum) pp. 96-137.

Tanigawa, T., Fuji, T. and Saegusa, T., 1998, A Developing of Distributed Learning Systems Using CORBA, In *IPSJ Journal*, Vol.39, No.7, pp.2380-2390.

Wakeman, L. and Jowerr, J., 1993, *PCTE: The standard for open repositories*, (Hertfordshire: Prentice Hall).

CHAPTER 9

Developing Domain Frameworks based on a Three-View Model

Takeo Hayase

Toshiba Corporation

ABSTRACT

This chapter describes developing domain frameworks based on a three-view model. This model consists of a domain analysis view, a layer view, and a mechanism view for building a framework that has high reusability, portability, and maintainability, respectively. The domain analysis view is used to extract domain-specific objects based on the modeling policy that we call Domain Reference Models (DRM). The layer view is used to divide a framework into three layers that are piled up vertically: an infrastructure layer, a generic layer, and a domain layer. The mechanism view is used to decide which the mechanism of whitebox frameworks or blackbox frameworks. We applied our approach to the framework development for automatic gates monitoring systems. We have a prospect of withdrawing the development costs of the framework by approximately five times of application deployment, and estimate approximately one hundred times of application deployment. Therefore, it is effective for software engineers to develop a framework based on the three-view model

9.1 INTRODUCTION

An object-oriented framework has been recently applied to software development as a reuse technique. A framework is a reusable design of all or part of a system that is represented by a set of abstract classes and the way their instances interaction (Fayad *et al.*, 1999). Many studies report that software productivity is greatly improved with frameworks. For example, several studies report frameworks that are generally used such as graphical user interface (GUI) (Schmucker, 1986) (Linton *et al.*, 1988) and operating system (Lewis *et al.*, 1993). Other studies report frameworks for a specific domain (Franek *et al.*, 1998) (Alfredsen *et al.*, 1998). In most of the previous studies, however, building a framework left to the individual skill of designers without engineering perspective. Moreover, software engineers must discuss various quality factors of software, such as reusability, maintainability, performance issues, and so on. This paper describes a three-view model for developing object-oriented frameworks. We give a new methodology based on the three-view model, and clarify an engineering process of building frameworks. This model consists of a domain analysis view, a layer view, and a mechanism view. Moreover, we demonstrate its effectiveness using an example of practical industrial systems.

In Section 9.2, we briefly survey benefits and problems of frameworks. In Section 9.3, we clarify the definition of a domain. In Section 9.4, we propose a new methodology based on a three-view model. In Section 9.5, we apply our approach to framework development for an industrial monitoring system. In Section 9.6, we demonstrate its effectiveness in this framework development. In Section 9.7, we compare our approach with related works. In section 9.8, we conclude this chapter.

9.2 EFFECTS AND PROBLEMS OF OBJECT-ORIENTED FRAMEWORKS

We briefly survey effects and problems of object-oriented framework.

9.2.1 Effects

By using frameworks, software engineers can especially improve productivity, reusability, and quality.

A framework is semi-complete application that can be specialized to produce custom applications. Because software engineers implement only the part of an application that must be customized to application-specific requirements, they can improve productivity.

If frameworks are used, the range of reusable components is limited to a class. Reusable components are the class structure including software architecture. Therefore, the range of reusable components can be wide in an application.

If a framework is often used, quality of the framework can be improved. Therefore, quality of an application can be improved.

9.2.2 Problems

Building a framework left to the individual skill of designers without engineering perspective. We focus on the difficulties of building the framework.

Software engineers must discuss system specifications of a domain to build the framework. The previous object-oriented methodologies (Rumbaugh, 1991) (Booch, 1994) (Jacobson, 1992) focus on phases of object-oriented analysis and design. In practical development, it is important for software engineers to discuss the static and dynamic aspect of an application at system-level, such as domain analysis and system analysis. However, the previous methodologies sufficiently have not had the viewpoint of system level.

When software engineers specify the frozen spots and hot spots by domain analysis, they focus on mainly reusability and extensibility. However, they must discuss various quality factors of software from these factors to performance issues and restriction on implementation. In most of the previous methodologies, this point has not been discussed sufficiently.

After a framework is applied, there is possibility that software engineers revise the framework because of new system specifications, and so on. The classification according to the mechanism of frameworks divides frameworks into whitebox frameworks and blackbox frameworks (Fayad, 1999). However, the guideline for deciding these mechanisms has not been discussed sufficiently.

9.3 DOMAIN

This section clarifies the definition of a domain. A domain is a set of many systems that are similar in a type of industry or business. These types are defined as follows.

- Type of industry

A type of industry represents a kind of industry or enterprise.

- Type of business

A type of business represents a kind of task on business.

We focus on the view of the type of industry. For example, Figure 9.1 shows the production domain, the control domain, the monitoring domain, the embedded domain, the accounting domain, the sales domain, the physical distribution domain, the broadcasting domain, and so on. On the other hand, a type of business is across several domains. For example, Figure 9.1 shows the transaction task, the stocking task, the planning task, and the event-monitoring task. In Figure 9.1, the production domain, the control domain, and the monitoring domain have the planning task, and so on. Then, the production domain, the control domain, the monitoring domain, and the embedded domain have the event-monitoring task. Furthermore, the sales domain, and the physical distribution domain have the stocking task. Then, the accounting domain, the sales domain, the physical distribution domain, and the broadcasting domain have the transaction task.

Based on these categories of domains, we divided these domains into two categories as shown in Figure 9.1. There are the manufacturing systems field, which have the control domain, the monitoring domain, and so on, and the information systems field, which have the accounting domain, the physical distribution domain, and so on.

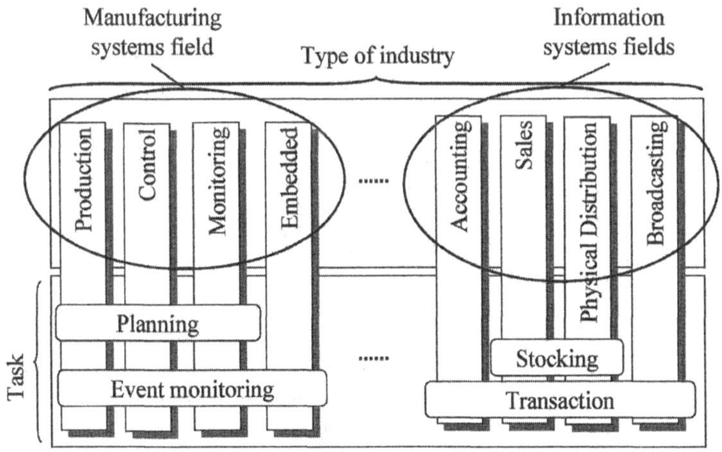

Figure 9.1 Definition of a domain

Figure 9.2 shows examples of systems in two system fields.

- Manufacturing systems field

These systems always have software that implements the processing of hardware. Without replacing hardware, these systems can easily modify system functions. As shown in Figure 9.2, there are many systems from large-scale software to medium-scale software.

- Information systems field

These systems always have software that implements the processing of businessperson. These systems perform the processing quickly and accurately. As shown in Figure 9.2, many systems always have medium-scale software.

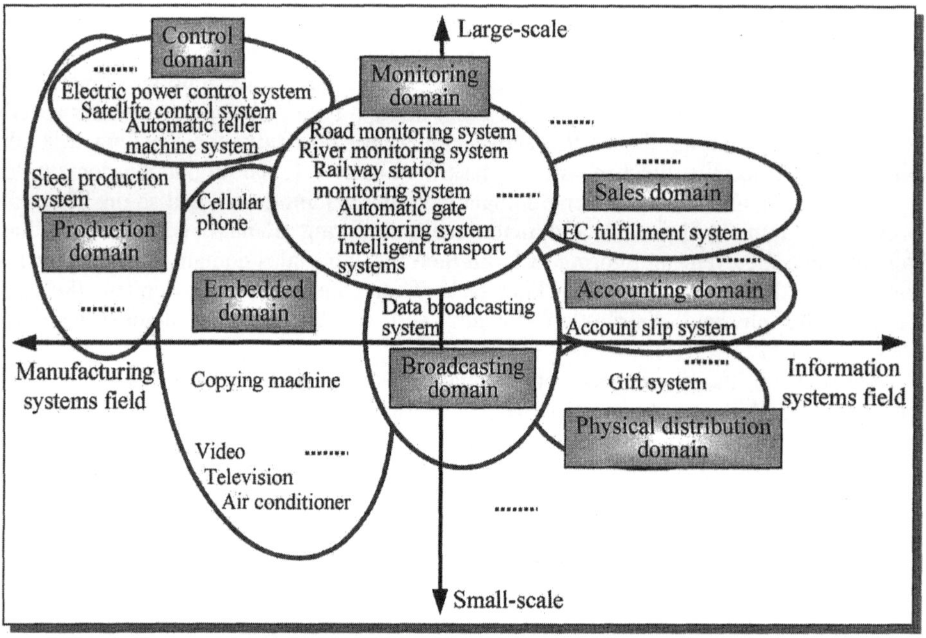

Figure 9.2 Systems in the manufacturing systems field and the information systems field

In this chapter, we focus on the control domain, the monitoring domain, and the embedded domain in the manufacturing systems field.

- Control domain

A system of the control domain control target nodes, and perform the control task. For example, this domain has electric power control systems, satellite control systems, automatic teller machine systems, and so on. As shown in Figure 9.3, this domain always has large-scale systems, and is required high performance.

- Monitoring domain

A system of the monitoring domain monitors target nodes, and perform the monitoring task. For example, this domain has railway station monitoring systems, road monitoring systems, river monitoring systems, and so on. As shown in Figure 9.3, this domain always has medium-scale systems, and is comparatively required high performance.

- Embedded domain

A system of the embedded domain has microprocessor, control target nodes, and perform the control task. For example, this domain has television, video, air conditioner, cellular phone, and so on. As shown in Figure 9.3, this domain always has many systems from large-scale software to medium-scale software.

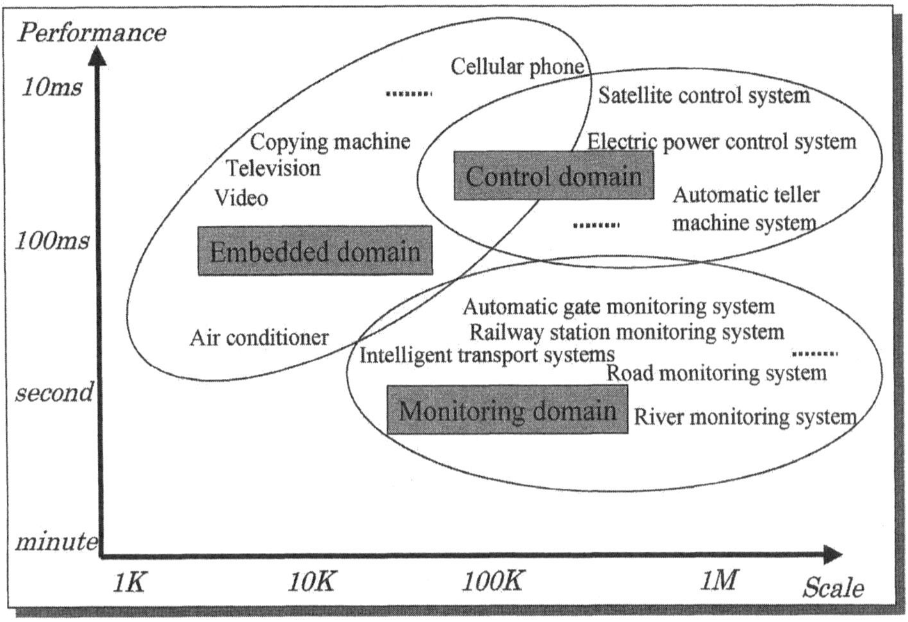

Figure 9.3 Systems in the manufacturing systems filed

9.4 THREE-VIEW MODEL FOR DEVELOPING DOMAIN FRAMEWORKS

We propose a new methodology based on a three-view model for developing an object-oriented framework. This model consists of a domain analysis view, a layer view, and a mechanism view as shown in Figure 9.4. This model has many-sided aspects of a framework that are necessarily not orthogonal. Software engineers construct analysis and design models based on these views. Our methodology focuses on frameworks that are used for domain-specific, and are not used for general purpose such as GUI. The domain-specific framework has many times of application deployment. This model can lead software engineers to frameworks that have high reusability, portability, and maintainability. In this section, we describe an outline of engineering processes of building the framework.

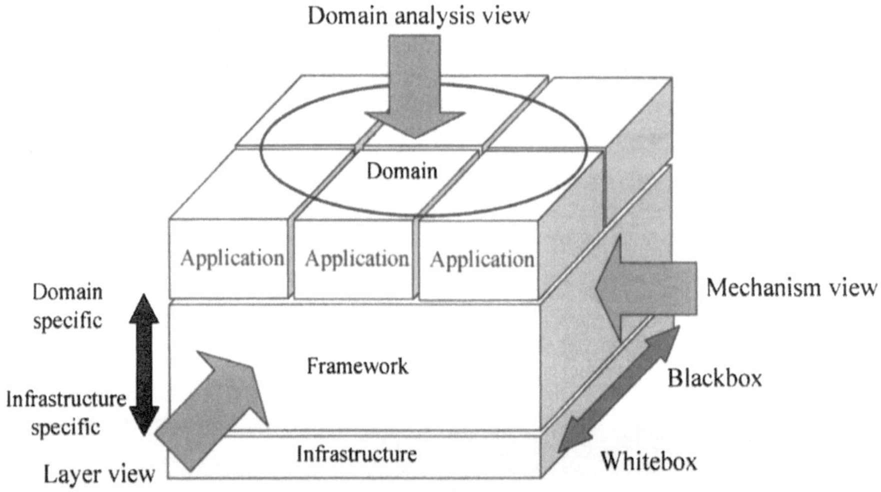

Figure 9.4 Viewpoint of framework development

9.4.1 Domain Analysis View

The domain analysis view focuses on reusability of a framework. Software engineers must sufficiently analyze a domain, and construct an object model to build the framework. In domain analysis, they analyze all information of the domain and domain knowledge. They specify both the frozen spots, which are common to all applications, and the hot spots, which are customized to system specifications (Pree, 1994). However, the previous methodologies (Rumbaugh, 1991) (Booch, 1994) (Jacobson, 1992) have not sufficiently supported the phase of domain analysis, and have not clarified how to extract objects and to use domain knowledge. Therefore, we propose a new concept that we call domain reference models (DRM), which are referred as the modeling policy when software engineers model domain-specific objects.

9.4.1.1 Definition of a DRM

A DRM is basic model for modeling domain-specific objects. This can improve reusability of software architecture in the same domain. In other words, this defines a general term for domain-specific objects that has a similar role in the same domain. Many systems in the same domain have similar features. Therefore, by defining software architecture based on a DRM, the range of reuse spread from single system to a domain that has similar many systems.

A DRM is referred as the modeling policy when software engineers model domain-specific objects in a domain. Therefore, we call domain reference models (DRM). A DRM should be defined in each domain. Domain analysis is needed to build a framework. Software engineers analyze and model features of a domain by domain analysis. To use all information of a domain effectively, they should construct a DRM in each domain. A DRM satisfy conditions as follows.

- To represent the static aspect of a domain independent of infrastructure
- To represent the dynamic aspect of a domain as message sequences between typical objects based on a DRM

- To clarify the guideline of object extraction and classification
- To represent as stereotypes of UML (Booch, 1998) in phases of object analysis and design

9.4.1.2 An example of DRM

We constructed various DRM according to target domains. For example, we constructed the Presentation-Entity-Relay-Service (PERS) model for an industrial monitoring domain. Many monitoring systems in this domain have several features as follows.

- Require target nodes, such as cameras, gates, shutters, sensors, and so on, to perform monitoring services or set the processing modes of them
- Obtain the processing status of them or the monitoring data, such as binary data, image data, video data, and so on
- Have GUI presentation
- Have many services that remotely control target nodes
- Send/receive data to/from target nodes
- Have several different communication protocols for communicating between monitoring systems and target nodes.

The PERS model is shown in Figure 9.5. The arrows show the relationship of "uses". A component of the origin uses a component of the destination.

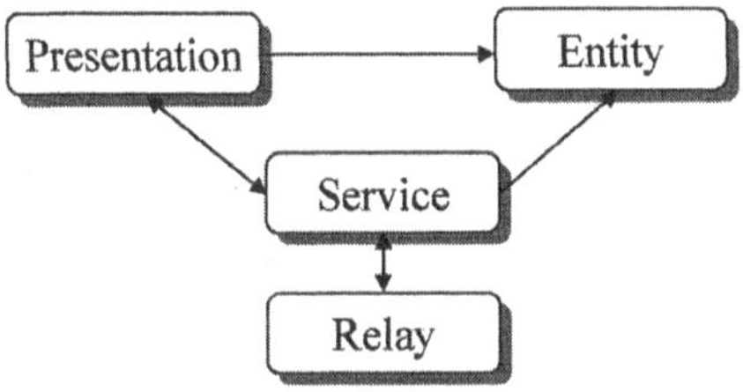

Figure 9.5 PERS (Presentation-Entity-Relay-Service) model

- Presentation

A *presentation* represents all information to a user by GUI.

- Entity

An *entity* contains data, which are processed in a domain.

- Relay

A *relay* communicates with target nodes and encapsulates a communication protocol.

- Service

A *service* contains the functional core of a domain, presents all information of the *entity* to the *presentation*, and sends control messages for target nodes to the *relay*.

9.4.1.3 How to construct DRM

If a domain do not have the established DRM, software engineers construct a DRM as follows. This activity is performed in the phase of preparation before they perform object-oriented analysis, and is one of activities in the phase of system analysis. After they define the DRM and clarify the guideline of object extraction and classification, they start activities in the phases of object-oriented analysis and design.

(1) Extraction of initial objects
Software engineers extract domain-specific objects for performing use cases that are analyzed in use case analysis. At this time, it is not necessary to extract all domain-specific objects completely. Therefore, they may extract domain-specific objects that perform typical basic use cases. Furthermore, they analyze system behavior according to use cases, and decide the role of each object.

(2) Classification of initial objects
Software engineers classify initial objects according to the role of each object. At this time, they decide the classification name of objects, and can represent it as stereotypes of class diagram.

(3) Modeling the DRM
If object extraction and classification are fixed, software engineers model system structure as the DRM. This is the basic model for object-oriented analysis.

9.4.1.4 Related reuse technique

DRM is one of reuse technique. Other reuse techniques are domain models, analysis patterns (Fowler, 1997), design patterns (Gamma, 1994), and frameworks. These are shown at two viewpoints of development processes and abstractness in Figure 9.6. Domain models are used to model domain knowledge, and are one of reuse technique at system level. DRM is one of reuse technique at system level, and is used at the phase of system analysis. Analysis patterns, design patterns, and frameworks are reuse technique at object level. Analysis patterns and design patterns are respectively used at the phase of object-oriented analysis and design. Frameworks are used at the phase of object-oriented design and implementation.

A DRM is important in domain-specific object modeling. Software engineers extract domain-specific objects based on the DRM, and implement a framework using analysis patterns and design patterns. Figure 9.7 shows the relationship between DRM and frameworks in class diagram. A domain has a DRM. In Figure 9.7, "Domain reference model A", "Domain reference model B", and "Domain reference model C" are DRM. Furthermore, software engineers model domain-specific objects based on the DRM. They implement these objects as a framework. Figure 9.7 shows that "Framework B1", "Framework B2", and "Framework Bn" are implemented based on "Domain reference model B". Furthermore, software engineers develop applications using the framework. Figure 9.7 shows that "Application B2 (a)", "Application B2 (b)", and "Application B2 (x)" use "Framework B2".

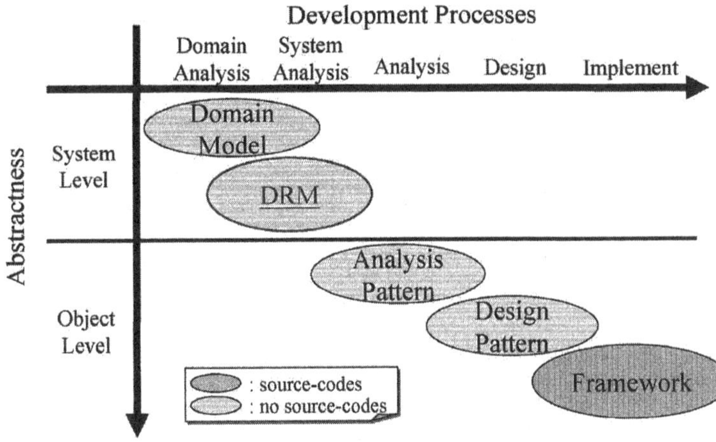

Figure 9.6 Domain Reference Models (DRM)

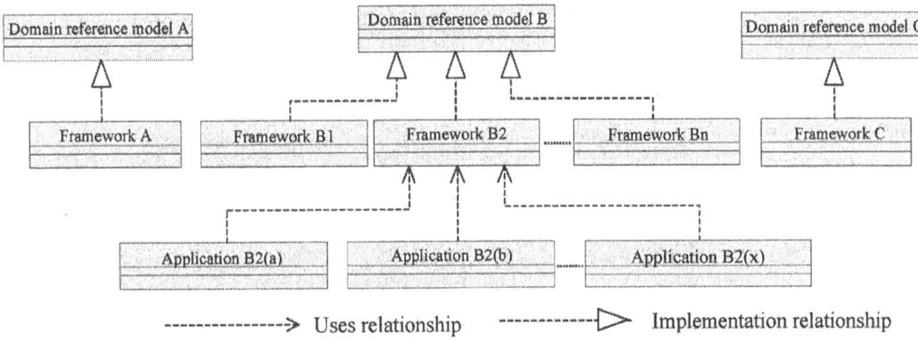

Figure 9.7 Relationship between DRM and frameworks

9.4.2 Layer View

The layer view focuses on portability of a framework. Software engineers sometimes change platform, and replace a part of the framework for restriction on implementation. This view is used to divide a framework into three layers that are piled up vertically: an infrastructure layer, a generic layer, and a domain layer. Each layer is defined in Table 9.1. Software engineers can deal with several quality factors of software, the change of infrastructure, and so on. For example, they focus on reusability and extensibility in early model of a framework. Performance of an application often depends on using the infrastructure efficiently. Therefore, if software engineers refine only the infrastructure layer of the framework, they can always satisfy performance issues.

Furthermore, we classify each layer into several categories in Figure 9.8. If the systematic architecture of frameworks is defined, software engineers can flexibly replace a part of the framework. Figure 9.8 shows just the systematic architecture of frameworks. This architecture does not mean that all applications use the same framework. For example, a framework in the *database* category of the infrastructure layer is not necessarily used in all domains. The infrastructure layer just has the *database* category.

Table 9.1 Layer of a framework

Layer	Role	Important quality factors
Infrastructure	Encapsulate infrastructure	Performance, genericity, and so on
Generic	Use for general domain	genericity, reusability, and so on
Domain	Use for specific domain	reusability, extensibility, and so on

9.4.2.1 Infrastructure layer

The infrastructure layer encapsulates infrastructure such as middleware, OS, and so on. This layer is classified into five categories that are *database*, *server logic*, *communication*, *system management*, and *security*.

- Database

The *database* category encapsulates databases such as an object-oriented database, a relational database, a file system, and so on.

- Server Logic

The *server logic* category encapsulates logic process on Web servers, and application servers.

- Communication

The *communication* category encapsulates communicational mechanisms such as Common Object Request Broker (CORBA), Remote Method Invocation (RMI), and so on.

- System management

The *system management* category has functions on information management of system configuration such as the composition information of computer nodes, for example, the relationship between clients and servers.

- Security

The *security* category has functions on system security such as certification of users, access control, and so on.

9.4.2.2 Generic layer

The generic layer includes functions independent of a domain. This layer is classified into three categories that are *presentation*, *data*, and *mechanism*.

- Presentation

The *presentation* category includes presentation formats of an application such as GUI.

- Data

The *data* category includes data formats such as data translation.

- Mechanism

The "*mechanism*" category includes behaviour of an application such as state transition mechanism, messaging between objects, and so on.

9.4.2.3 Domain layer

The domain layer includes functions for domain-specific. This layer is classified several categories that are *production*, *control*, *monitoring*, *physical distribution*, *broadcasting*, and so on. For example, the *control* category and the *monitoring* category show as follows.

- Control

The *control* category includes a domain that controls hardware units such as card unit, gates, shutters, and so on.

- Monitoring

The *monitoring* category includes a domain that monitors target nodes such as cameras, gates, shutters, and sensors, and so on.

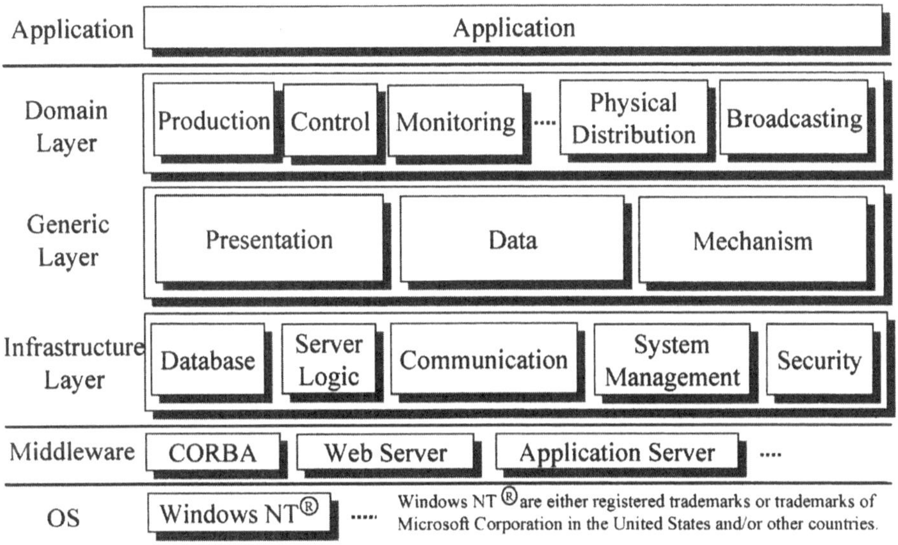

Figure 9.8 Architecture of layered frameworks

9.4.3 Mechanism View

The mechanism view focuses on maintainability of a framework. Frameworks are categorized into whitebox frameworks and blackbox frameworks (Fayad, 1999). Building a framework requires software engineers to be able to maintain it according to its purpose. Therefore, it is important for them to decide the appropriate mechanism of frameworks. The previous studies have already discussed its features and mechanisms (Fayad, 1999) as shown in Table 9.2.

A whitebox framework has the inheritance mechanism. If software engineers develop an application, they derive application-specific classes from base classes of the framework. It is easy for them to build the framework, but is difficult for them to use it.

On the other hand, a blackbox framework has the composition mechanism. If software engineers develop an application, they implement application-specific classes that have interface of classes in the framework. It is difficult for them to build the framework, but is easy for them to use it.

Table 9.2 Mechanism of a whitebox framework and a blackbox framework

Category	Mechanism	Way of use	Building	Using
Whitebox	Inheritance	Derive base classes	Easy	Difficult
Blackbox	Composition	Implement interface	Difficult	Easy

It is difficult for software engineers to obtain sufficient effects using a framework without understanding features of both mechanisms. However, the previous studies (Pree, 1994) (Schmid, 1996) have not sufficiently discussed the guideline to decide these mechanisms. Therefore, we propose the guideline using the previous two views, which are a domain analysis view and a layer view in the three-view model, in Figure 9.9.

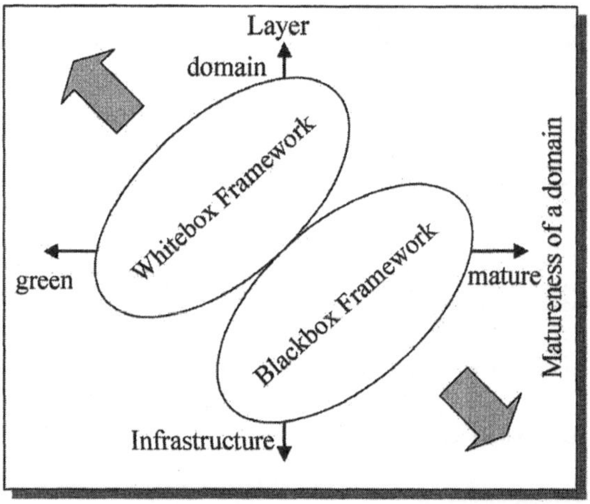

Figure 9.9 Guideline for deciding mechanisms of the framework

9.4.3.1 Matureness of a domain

The first point is matureness of a domain. This depends on stability of system specifications. A whitebox framework is suited for a green domain, and a blackbox framework is suited for a mature domain as shown in Figure 9.9.

In the green domain, software engineers must often customize functions of an application. In the whitebox framework, established functionality of the framework is extended by inheriting from base classes of the framework, and override predefined hook methods. Therefore, the whitebox framework is suited for the green domain.

On the other hand, in the mature domain, software engineers require the framework to be more rigid in the domain. In the blackbox framework, established functionality of the framework is extended by defining components that conform to particular interfaces, and integrating components into the framework. Therefore, the blackbox framework is suited for the mature domain.

9.4.3.2 Target layer

The second point is a target layer: an infrastructure layer, a generic layer, and a domain layer. A blackbox framework is suited for the infrastructure layer and generic layer, and a whitebox framework is suited for the domain layer as shown in Figure 9.9.

The infrastructure layer and generic layer are often used as the framework for general-purpose. This framework must be more rigid. In the whitebox framework, exposing framework codes to software engineers may lead to some problems. On the contrary, the blackbox framework supports stable interfaces and functions. Therefore, the blackbox framework is suited for the infrastructure layer and generic layer.

On the other hand, the domain layer is used as the framework for a specific domain. It is difficult for software engineers to specify how system specifications are changed in the near future. If these are changed, they must refine the framework. To use the blackbox framework leads to lack of flexibility. Therefore, the whitebox framework is suited for the domain layer.

9.4.4 Building Frameworks using Three-View Model

We describe an outline of processes of building a framework using the three-view model. Software engineers focus on the domain analysis view, the layer view, and the mechanism view, respectively. The outline of the processes of building the framework is as follows.

9.4.4.1 Domain analysis

The aim of domain analysis is to survey all domain knowledge and information in a domain.

Process1: Domain modeling
After software engineers survey domain knowledge and information, and discuss system specifications in the domain, they build a domain model. Especially, it is important for them to clarify similarity, commonality, and difference of many systems in the domain.

9.4.4.2 System analysis

The aim of system analysis is to define the structure and behavior of a system from the viewpoint of users.

Process2: Use case analysis
Software engineers decide the range of system development, and define system specifications. They analysis use cases, which are functions of the system from the viewpoint of users, and interactions between this system and actors, which are users or external systems.

Process3: System architecture decision
Software engineers decide the system architecture from the viewpoint of users. For example, they determine hardware, middleware, programming language, and so on. Furthermore, they survey reusable software components.

Process4: DRM definition
Software engineers define DRM based on domain analysis. They can reuse the established DRM, or extend it if they need. If the established DRM are not appropriate, they construct a new DRM as shown in Section 9.4.1.3.

9.4.4.3 Object-oriented analysis

The aim of object-oriented analysis is to extract domain-specific objects and build the structure and behavior of these objects.

Process5: Object-oriented analysis
The model based on the DRM show the guideline of object extraction and classification. They construct domain-specific objects based on the DRM.

Process6: Hot spot analysis
Software engineers specify both the frozen spots, which are common to all applications, and the hot spots, which are customized to system specifications.

9.4.4.4 Object-oriented design

The aim of object-oriented design is to clarify how to implement the object model of object-oriented analysis.

Process7: Architecture decision
Software engineers decide the framework architecture. The architecture represents layers of the framework and categories in each layer as shown in Figure 9.8.

Process8: Mechanism decision
Software engineers decide the mechanism of each category. This mechanism decision represents which they choose the whitebox framework or the blackbox framework.

Process9: Framework design
Software engineers discuss the object model of the framework to deal with the hot spots, and design static models and dynamic models.

9.4.4.5 Object-oriented programming

The aim of object-oriented programming is to implement software based on object design.

Process10: Layered framework development
Software engineers develop the framework of each layer. In Figure 9.10, there are three development phases for building the infrastructure layer framework, the generic layer framework, and the domain layer framework. Furthermore, software engineers vertically pile up from the infrastructure layer framework to the domain layer framework as shown in Figure 9.10.

Process11: Framework refinement
Software engineers refine the whole framework in order to satisfy all system specifications. For example, they focus on reusability and extensibility of a framework for a system at first. However, this framework does not satisfy the performance issues of the system. At this time, they can replace the infrastructure layer framework with new infrastructure layer framework, which is focus on the performance issues. Then, they can replace other layer frameworks with new layer frameworks. The points of replacement in each layer are as follows.

- Infrastructure layer framework

In the infrastructure layer framework, software engineers discuss efficiency for using infrastructure resources. For example, the issue how to implement the structure that connect an object with a thread, which is a unit of concurrent processing, is concerned with this layer framework.

- Generic layer framework

In the generic layer framework, software engineers discuss variation for implementing generic functions. For example, there are several mechanisms for state transition. The simplest mechanism is the way using a state variable. The other mechanism is the way using the state pattern, which is one of design patterns (Gamma, 1994).

- Domain layer framework

In the domain layer framework, software engineers always discuss system specifications. From the viewpoint of system functions, the unique object model is enough. On the other hand, when they refine the infrastructure layer framework and the generic layer framework, there is a possibility that the whole framework does not satisfy all system specifications. Therefore, it is effective that they discuss alternatives in this layer framework.

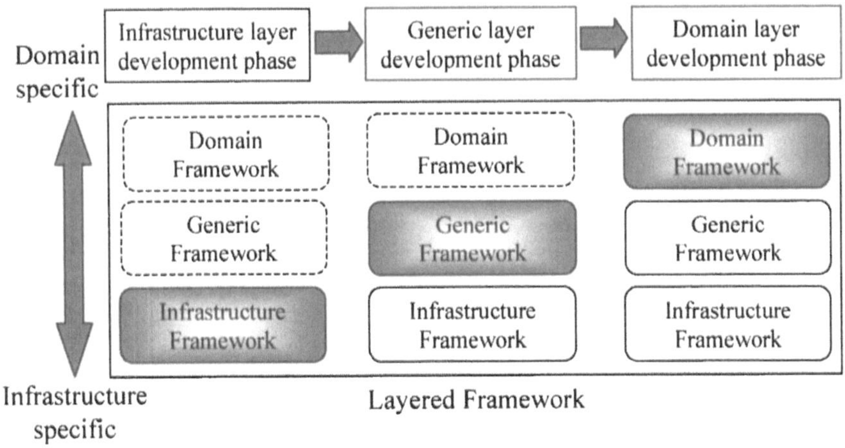

Figure 9.10 Development process of the layered framework

9.5 AN EXAMPLE OF INDUSTRIAL MONITORING SYSTEMS

We applied our approach as shown in Section 9.4 to the framework development for automatic gate monitoring systems.

9.5.1 Target system

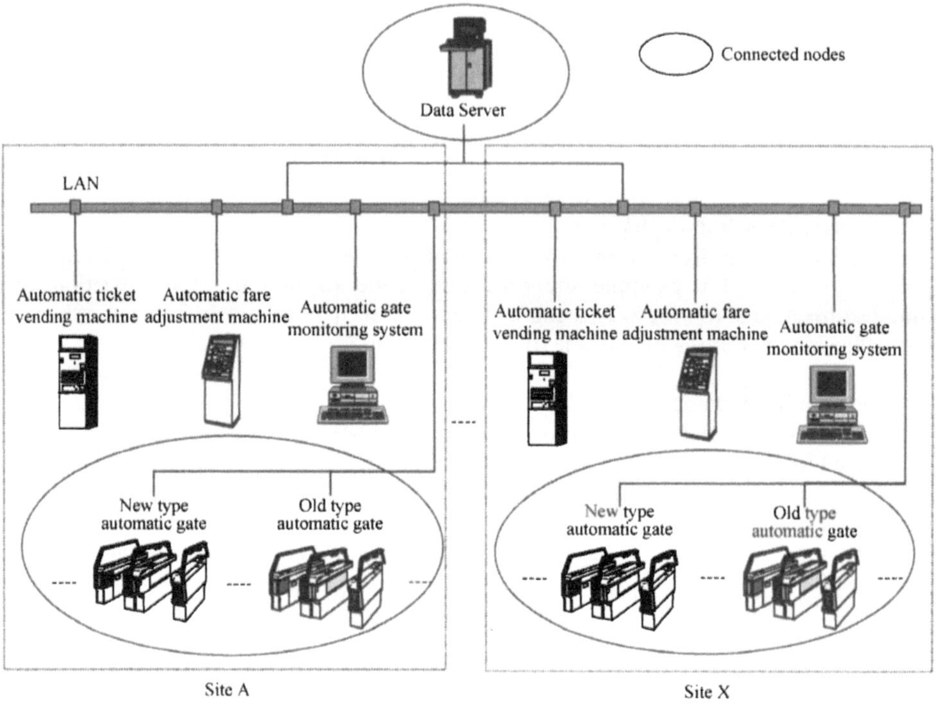

Figure 9.11 Automatic gate monitoring systems

An automatic gate monitoring system is an industrial monitoring system to use as a social infrastructure at railway stations. As shown in Figure 9.11, this system connects with a data-server, and new or old-type automatic gates by Local Area Network (LAN). This system is installed in each group that is gathered automatic gates, and connects with a data-server. This system has approximately fifty services as monitoring tasks. The size of source codes using C++ programming language is approximately one hundred thousand steps. In monitoring tasks, communication data between this system and an automatic gate is numerical binary data. For example, the processing status of an automatic gate is defined by receiving data from the automatic gate. Moreover, collecting data from the automatic gate shows historical information of providing users services. This system receives numerical binary data from the automatic gate. Furthermore, the size of communication data is several hundreds K-byte. However, this data is frequently communicated at an interval from several seconds to several ten seconds. The services of this system are briefly classified as follows.

- Display the processing status of automatic gates, and set the processing modes of them.
- Collect data from automatic gates, calculate a total of this data, and send calculated data to the data-server.
- Receive the setting parameters for automatic gates from the data-server, and distribute them to each automatic gate.
- Display the processing status of communication between this system and automatic gates, or between this system and the data-server.

9.5.2 Framework Development Processes

We applied our approach to automatic gate monitoring systems. However, we had developed several applications by using object-oriented approach and no using the framework for this system before we have developed applications by using the framework for this system. We could sufficiently refine the object model, and reuse the previous results in the phases from domain analysis to object-oriented analysis. The activities in the phases from domain analysis to object-oriented design are shown as follows.

1.5.2.1 Domain Analysis

Process1: Domain modeling
Domain Analysis was proceeded at two points of understanding this domain and surveying system specifications. It is possible to install new-type automatic gates, and to add some services. Because system specifications are frequently changed, an application must often be customized. Therefore, this domain is green. Furthermore, the main system requirements are shown as follows. In particular, reusability and extensibility are important in the following quality factors of software. We must develop many applications, and be required short lead-time by end-users.

- Reusability

System specifications vary with each application, and are often added or changed. If we separately develop the application for each end-user, development costs increase. Therefore, reusable components are required.

- Extensibility

It is difficult for software engineers to predict new required specifications. For example, new services or new operating rules are added. Therefore, this domain is green. Software engineers must rapidly develop many systems.

- Performance

This system must communicate numerical binary data with automatic gates within several hundreds milliseconds. Furthermore, this system must calculate approximately ten thousands data within ten minutes.

9.5.2.2 System analysis

Process2: Use case analysis
There are difference according to system configurations and specifications. We clarify all use cases of this system as shown in Figure 9.12.

Process3: System architecture decision
We adopt C++ programming language because of efficiency, and apply component technology to GUI.

Process4: DRM Definition
This system is an application for the industrial monitoring domain. We used the PERS model as shown in Section 9.4.1.2. The reason of using this model is as follows.

- Services are provided by co-operating with automatic gates.
- System specifications of services are changed frequently.
- Communication protocols vary with new-type or old-type automatic gates.

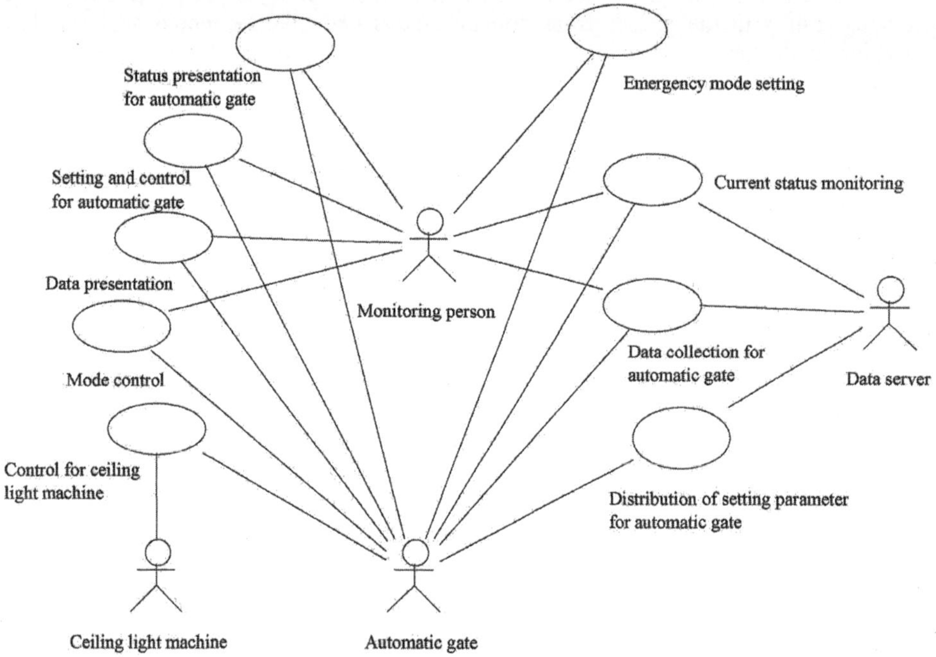

Figure 9.12 Use case diagram

9.5.2.3 Object-Oriented Analysis

Process5: Object-oriented analysis
We extract domain-specific objects based on the PERS model as follows.

- Objects for GUI presentation

An object in the *presentation* component is extracted from the viewpoint of GUI components. For example, an object in this component is the "setting-mode GUI" object.

- Objects for information in the domain

An object in the *entity* component is extracted from the viewpoint of processed units in this system. For example, an object in this component is the "setting-data" object.

- Objects for communicating with automatic gates

An object in the *relay* component is extracted from the viewpoint of automatic gates. For example, an object in this component is the "old-type automatic gate" object.

- Objects for servicing monitoring tasks

An object in the *service* component is extracted from the viewpoint of services in this system. For example, an object in this component is the "monitoring service" object.

Process6: Hot Spot Analysis
We developed several applications using object-oriented technology, and examined specifications of the framework for this system. The hot spots are classified into the PERS model as shown in Table 9.3. Most of the hot spots are concentrated on the *service* component. Some points of the hot spots are briefly as follows.

- A button is added by adding processed data in this system. This is related to the *presentation* component.
- Error information of monitoring data varies with connected automatic gates. This is related to the *entity* component.
- Communication protocols vary with connected automatic gates. This is related to the *relay* component.
- Provided services vary with connected data-server. This is related to the *service* component.

Table 9.3 Ratio of Hot Spots in the PERS Model

Component	Ratio (%)
Presentation	20
Entity	20
Relay	10
Service	50

9.5.2.4 Object-oriented design

Process7: Architecture decision
We specified layers and categories in each layer. The architecture of this framework is shown in Figure 9.13. A shadowed square shows a category of frameworks. A shadowed and round-corner square shows a framework. The infrastructure layer framework has two categories: *Database* and *Communication*. The *Database* category has components for accessing to common files or memories. The *Communication* category has components for communicating data between processes in Windows NT. The generic layer framework has two categories: *Presentation* and *Mechanism*. The *Presentation* category has components for consisting of monitoring GUI. The *Mechanism* category has components for communicating messages between objects, and accessing data objects. The domain layer framework has domain-specific objects in the PERS model.

Process8: Mechanism decision

We implemented the framework as the blackbox framework for the *presentation* component, and the whitebox framework for other components in Figure 9.13. The *Presentation* component is implemented using component technology because of having GUI. Therefore, we implemented this component as the blackbox framework. Because other components to be customized application frequently, we implemented other components as the whitebox framework.

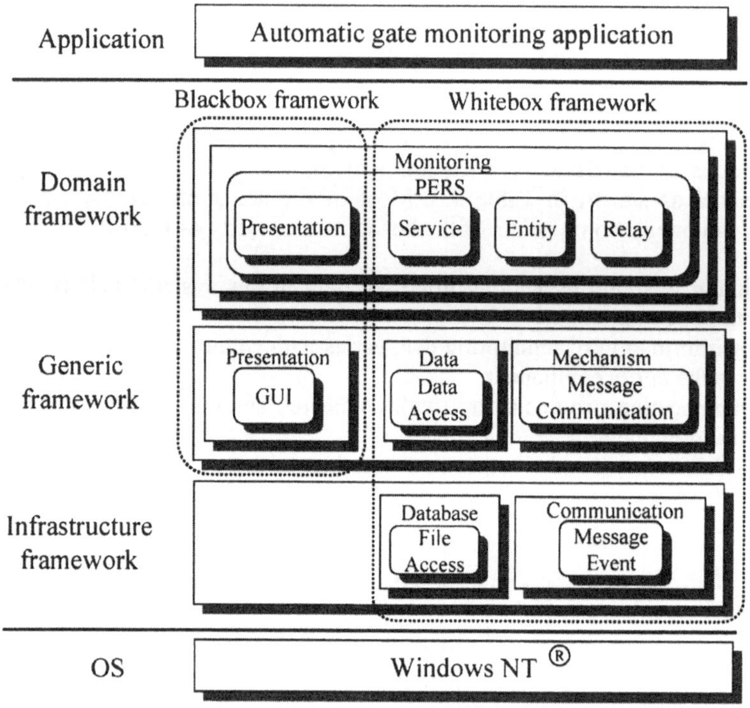

Figure9.13 Architecture of automatic gate monitoring systems

Process9: Framework Design

In this system, the *presentation* component was implemented as the blackbox framework, and other components were implemented as the whitebox framework. In this framework, monitoring services are performed by the mechanism of message communication between domain-specific objects. This mechanism was designed as the *Mechanism* category in the generic layer framework. The *Mechanism* category in the generic layer framework is related to the *Communication* category in the infrastructure layer framework. By hot spot analysis, we designed domain-specific objects that are related to the hot spots.

For example, we explain the *service* component in the domain layer framework. The objects in this component are divided into the *Manager* objects and the *Data Accessor* objects. The *Manager* objects have a role to perform monitoring services. The *Data Accessor* objects have a role to access to the *entity* component, and provide processed data in this system by using the *Mechanism* category in the generic layer framework. We designed the framework for this system, which has an architecture that the domain layer framework using the PERS model is related to the generic and the infrastructure layer framework.

9.6 EVALUATION

We evaluated the three-view model approach through developing the framework for automatic gate monitoring system. The development of some products is currently proceeding to use the framework. We evaluated effects for using framework and merits based on each view in the three-view model.

9.6.1 Effects for using framework

Reusability and extensibility are important in this system. As a way to evaluate these quality factors, we compared development costs in the prototype application using frameworks based on our approach with them in the product application using no frameworks. We assume typical modification of system specifications as follows.

- Adding services for the data server

This implement functions for the data server. We assume system specifications that are to collect numerical binary data from automatic gates and to send these data to the data server.

- Adding destination of status notification

We assume that this system not only send status information to monitor of this system, but also do it to the data server.

- Adding interphone functions

We assume that this system is added interphone functions to notify that automatic gates are unusual state.

In the above modification of system specifications, we evaluate development costs. In the first deployment, because we must development the framework, we need to evaluate our approach including development costs of the framework. Therefore, we decide what times of application deployment we can withdraw development costs of the framework by the following formula.

$$DfwDevCost < n\ (ObjAplDevCost - DfwAplDevCost)$$

DfwDevCost is development costs of the framework. *N* are times of application deployment. *ObjAPlDevCost* is development costs of an application using no framework. *DfwAPlDevCost* is development costs of an application using the framework. The result of this evaluation is shown in Figure 9.14. We could reduce development costs using the framework to 90% of them using no framework. Furthermore, we have a prospect of withdrawing development costs of the framework by five or six times of application deployment. In this application, we estimate approximately one hundred times of application deployment. Therefore, it is effective for software engineers to develop the framework based on the three-view model.

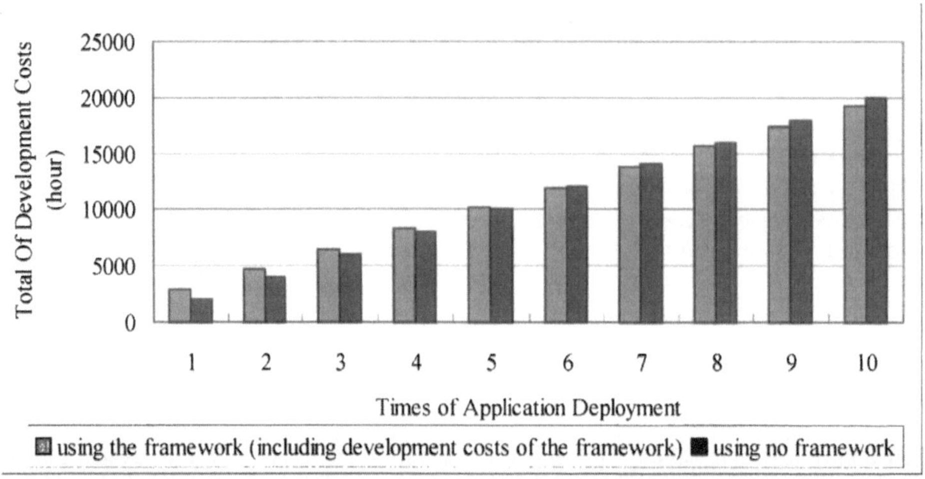

Figure 9.14 Comparison of Development Costs

9.6.2 Domain Analysis View

The core technique in the domain analysis view is a new concept that we call DRM. In the development of industrial monitoring systems, we use the PERS model for the monitoring domain. We describe its effectiveness for using this model.

For example, automatic gate monitoring systems connect with old and new-type automatic gates. The service logic for collecting data from automatic gates and communication protocols vary with these gates. In case of old-type automatic gates. there is shown in Figure 9.15 (a). If this system requests an old-type automatic gate to collect data, and this gate sends data to this system. On the other hand, in case of new-type automatic gates, there is shown in Figure 9.15 (b). If this system requests a new-type automatic gate to collect data, this gate send a message to accept the request and it sends data to this system.

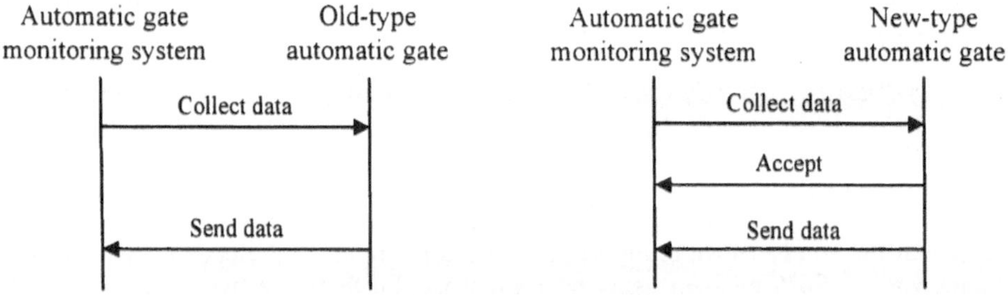

(a) In case of old-type automatic gate (b) In case of new-type automatic gate

Figure 9.15 Difference of service logic between old-type and new-type automatic gate

In this system, we deal with this issue by constructing the class structure in Figure 9.16. In this figure, we show the outline of class diagram, this system has approximately four hundred classes. This system has a function to collect processed data from automatic gates. The *relay* component in the PERS model has a role to communicate with automatic gates, and to encapsulate communication protocols. A *MonitoringService* object in the *service* component requests a *CollectingService* object in the *service* component to collect processed data. The *MonitoirngService* object use common interfaces of the *CollectingService* object independent of automatic gates. An *OldCollectingServiceImp* object and a *NewCollectingServiceImp* object, which are derived from a *CollectingServiceImp* object, handle service logic. Therefore, the *MonitoirngService* object is independent of the type of automatic gates. By using the PERS model, we localize the difference of the type of automatic gates in the *OldCollectingServiceImp* object and the *NewCollectingServiceImp* object. If we need other new-type automatic gates, we add a derived object from the *CollectingServiceImp* object. Therefore, we could improve reusability by using the PERS model.

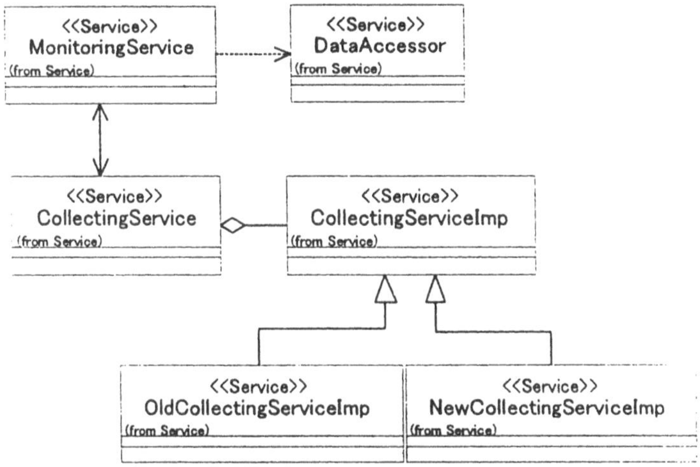

Figure 9.16 Outline of class diagram

9.6.3 Layer View

We divided the framework into three layers: an infrastructure layer, a generic layer, and a domain layer. We needed to redesign the *Database* category in infrastructure layer framework. By using the simulator for automatic gates, we made sure that this system can communicates numerical binary data, such as mode-setting data, with automatic gates within approximately several hundred milliseconds.

However, if this system receive many data from automatic gates, this system gradually slowed its speed of calculation. Then, we improved objects of the *File Access* category in the infrastructure framework. The objects of this category use the relational database. By tuning how to use this database, we deal with this performance issue. We can localize this issue in the specific part of the framework. Therefore, we could improve portability by using the layer view.

9.6.4 Mechanism View

We describe an evaluation from the viewpoint of a way to maintain the framework. We estimate approximately one hundred times of application deployment. System specifications vary with each application. In even the same application, system specifications are changed once several months. For example, new-type automatic gates are installed, and several services are added. Furthermore, we must often develop more than ten applications a half year.

Therefore, this domain is green. In this domain, because we implement the framework as the whitebox framework, except the *presentation* component, we can customize functions of an application according to system specifications. We could decide the appropriate mechanism of the framework. Therefore, we could improve maintainability by using the mechanism view.

9.7 RELATED WORKS

We compare our approach with related studies. We focus on the previous object-oriented methodologies, and the framework development methodologies.

Rational Unified Process (RUP) is one of the previous object-oriented methodologies. In this methodology, the "4+1" view model approach (Kruchten, 1995) is provided. This model composes of multiple views or perspectives. In order to address large and challenging architecture eventually, the model is made up of five views: a logical view, a process view, a physical view, a development view, and a use-cases view. This model focuses on the view of the various 'stakeholders' of the architecture: end-user, developers, system-engineers, project-managers, and so on. On the other hand, our approach is made up of three views: a domain analysis view, a layer view, and a mechanism view. We focus on quality factors of software: reusability, portability, and maintainability. The "4+1" view model and our model can harmonize mutually because of having different viewpoints.

As the framework development methodologies, there are hot-spot-driven approach (Pree, 1994), and hot-spot subsystems approach (Schmid, 1996), layering approach in large systems (Baumer, 1996), and Catalysis approach (D'Souza, 1998). These approaches focus on the view such as a domain analysis view or a layer view in our model. On the contrary, our approach focuses on three views. We compare our approach with the previous approaches from the viewpoint of each view.

In hot-spot-driven approach (Pree, 1994), which extend the previous object-oriented methodology, software engineers specify the frozen spots and hot spots, and use design patterns (Gamma, 1994). Hot-spot subsystems approach (Schmid, 1996) provided the structure of base classes to be suite for the hot spots, which is referred to as hot-spot subsystems. These two approaches focus on design of a framework, and point that domain analysis is important in order to specify the hot spots. However, these approaches do not focus on the issues how to deal with the results of domain analysis. Our approach provides a new concept that we call DRM. We show the results of domain analysis as basic model for modeling domain-specific objects. Our approach is different from these approaches at the focusing phase of the framework development. These approaches can support each other.

In layering approach in large systems (Baumer, 1996), frameworks in large systems are divided into five layers: *Application*, *Business Section*, *Business Domain*, *Desktop*, and *Technical Kernel*. This approach is similar to our approach from the viewpoint of layering frameworks. However, this approach relatively focuses on business applications. Our approach relatively focuses on industrial applications.

Catalysis approach (D'Souza, 1998) generalizes the concept of frameworks to object modeling and specification activities, permitting construction of specification and design models by composition. This approach emphasizes component-based development. It means using component technology, such as JavaBeans, Component Object Model (COM), and so on, as blackbox frameworks. Our approach provides the guideline to decide the appropriate mechanism of frameworks, which are categorized into whitebox frameworks and blackbox frameworks. Furthermore, Catalysis approach relatively focuses on business applications. Our approach relatively focuses on industrial applications.

9.8 CONCLUSIONS

This chapter described the three-view model for developing object-oriented frameworks. Furthermore, we applied our approach to the framework development for automatic gate monitoring systems in the monitoring domain, and evaluated our approach.

Our methodology is based on the three-view model. The three-view model consists of a domain analysis view, a layer view, and a mechanism view. This model can lead software engineers to a framework that has high reusability, portability, and maintainability. The domain analysis view is used to clarify all information and domain knowledge by using basic model that we call DRM, so that the framework has high reusability. The layer view is used to divide the framework into three layers that are piled up vertically: an infrastructure layer, a generic layer, and a domain layer, so that the framework has high portability. The mechanism view is used to decide which the mechanism of whitebox frameworks or blackbox frameworks, so that the framework has high maintainability. Furthermore, we applied our approach to the framework development for automatic gate monitoring systems in the monitoring domain. By developing a prototype application, we have a prospect of withdrawing development costs of the framework by five or six times of application deployment. In this application, we estimated approximately one hundred times of application deployment. Therefore, it was effective for software engineers to develop a framework based on the three-view model.

In the near future, we must discuss three problems as follows. First, we must evaluate more quantitative effects of our approach. It is difficult for software engineers to develop the same framework using different approach because of comparative conditions and the strategy of the development. Therefore, we plan to evaluate that what times of application deployment actually withdraw development costs of the framework. Next, we apply our approach to other applications in other domains. We may need to customize our approach in other domains. Finally, we need more refined guideline for developing the framework.

REFERENCES

Fayad, M. E., Schmidt, D. C. and Johnson, R. E., 1999, Building Application Frameworks, Addison-Wesley.

Schmucker, K. J., 1986, Object-Oriented Programming for the Macintosh, Hayden Book Company.

Linton, M. A., Calder, P. R. and Vlissides, J. M., 1988, InterViews: A C++ Graphical Interface Toolkit, Stanford University.
Lewis, T., and et al., 1993, Object-Oriented Application Framework, Manning Publications.
Franek, B. and Gaspar, C., 1998, SMI++ Object Oriented Framework for Designing and Implementing Distributed Control Systems, IEEE Transaction on Nuclear Science, Vol. 45, No. 4, pp. 1946–1950.
Alfredsen, K. and Saether, B., 1998, An Object-Oriented Framework for Water Resource Planning, Proceedings of the Thirty-First Hawaii International Conference on System Science, Vol. 7, pp. 441–450.
Rumbaugh, J., and et al., 1991, Object-Oriented Modeling & Design, Prentice Hall.
Booch, G., 1994, Object Analysis and Design with Applications. Second Edition., Benjamin Cummings Publishing Company.
Jacobson, I., and et al, 1992, Object-Oriented Software Engineering: A Use Case Driven Approach, The ACM press, A Division of the Association for Computing Machinery.
Booch, G., 1998, The Unified Modeling Language User Guide, Addison-Wesley.
Pree, W., 1994, Design Patterns for Object-Oriented Software Development, Addison-Wesley.
Fowler, M, 1997, Analysis Patterns: Reusable Object Models, Addison-Wesley.
Gamma, E., Helm, R., Johnson, R. and Vlissides, J., 1994, Design Patterns, Addison-Wesley.
Kruchten, P., 1995, Architectural Blueprints - The "4+1" View Model of Software Architecture, IEEE Software, Vol. 12, No. 6, pp. 42–50.
Schmid, H. A., 1996, Design patterns for constructing the hot spots of a manufacturing framework, Journal of Object-Oriented Programming, Vol. 9, No. 3, pp. 25–37.
Baumer, D., Knoll, R., Gryczan, G. and Zullighoven, H., 1996, Large scale object-oriented software-development in a banking environment, Object-Oriented Programming, 10th European Conference, IEEE, pp. 73–90.
D'Souza, D., and Wills, A. C.: Catalysis, 1998, Component and Framework Based Development, Addison-Wesley.

CHAPTER 10

Development of a Support Tool for Business Process Analysis

Automatic Generation of IDEF0 diagrams

Toyohiko Hirota

Kyushu Sangyo University

Satoshi Kumagai

Yamatake Corporation

ABSTRACT

To improve business processes, we have to understand the current processes in detail, but business analysis is not an easy task and takes considerable costs. We suppose that a business support system is constructed based on a business model. Therefore, in our research, we aim to construct a business model from data flows in a business support system. In this paper, we report a procedure to construct an IDEF0 diagram by analyzing data items in Web pages which are the interface of a business support system, and a prototype tool to generate the diagram automatically.

10.1 INTRODUCTION

To improve business processes, its current status should be captured in detail. This is often referred to as business process model. Knowledge about business process is often obtained through interviews with domain experts, or through collection and analysis of data used in the process. Those results should be organized as a comprehensive business process model, which makes it easy to understand the process. That model would facilitate communication among stakeholders of the process, and knowledge reuse would also be promoted (Itoh, *et al.*, 1998).

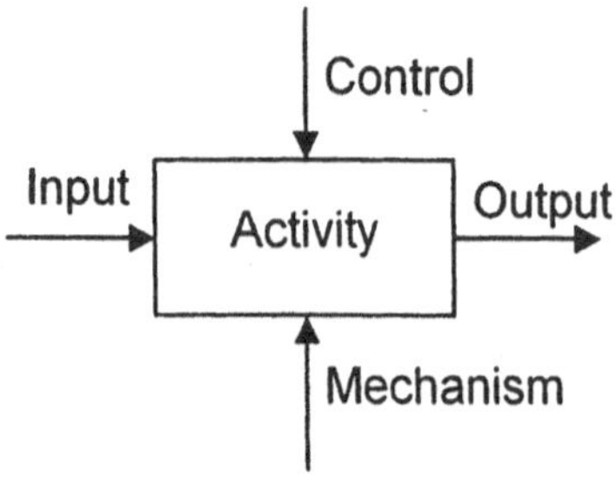

Figure 10.1 Element of IDEF0 diagram

Today, IDEF (Integration Definition Method) exists in practice as a large number of different technique as IDEF family of methods. Among those, IDEF0 (Marca *et al.*, 1988) has earned wide acceptance for use in BPR and system design and has been formalized as FIPS (Federal Information Processing Standards). In practices of system developments, it is not often the case that detailed model in IDEF0 for target business process is built. Actually, implicit model of business process is often made by individual developer and never specified to be shared or reused. This is typical particularly in small-scale system development or situation in which rapid adaptation for changing business process is required. Since it is difficult for system developers to take time for specifying the business model, they end up building implicit model in mind. which will never be specified nor reused, to advance the system development.

In this research, we assume that even if business process model is not created in a formal and explicit manner in the course of development, it should be reflected in the software as final product. We provide a method of reverse-engineering business model from the software product developed. That method enables to automatically generate the model from the product, and the obtained model would be useful for maintenance or enhancement of the system. Also, we recognize that might provide base for reengineering business process because that model should provide information about interaction between human and computer activities. We develop a prototype which automatically generates IDEF0 diagram as model form input/output data items in Web pages of the target system.

10.2 IDEF0

10.2.1 Notations

IDEF0 is a method of modeling process of system. Here, system means business and production activities. Process is a unit of collective activities. It includes not only activities by computers and machines activities but also those by human. Interactions among all of them are also specified. In IDEF0, system is represented as processes and their mutual relationships.

Main constructs of IDEF0 diagrams are activities and things associated to them. Each activity is represented as box in Figure 10.1. Things are classified into input, output, control and mechanism. They can be both material and information. Activity may

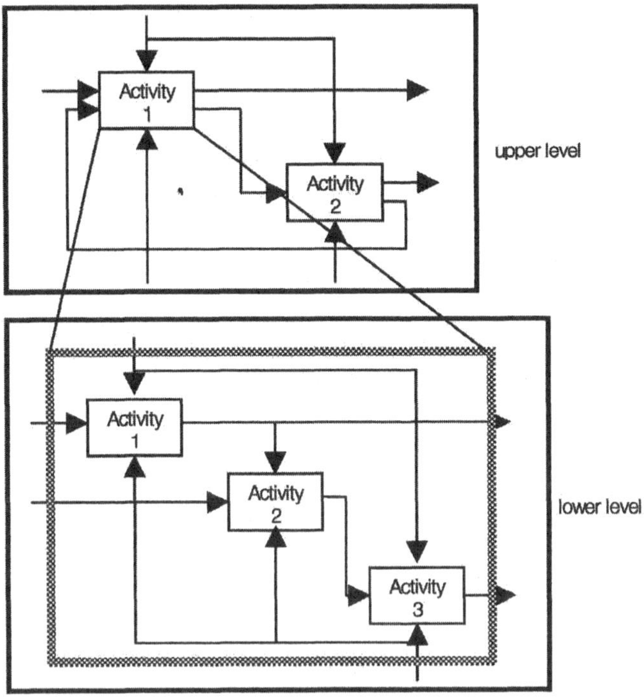

Figure 10.2 Hierarchy of IDEF0 diagram

use a mechanism which is based on the indication of a control application, and convert an input into an output. Here, input, control, output, and mechanism are collectively known as ICOMs. They may include data and information plus anything else which can be described in a business process (e.g. schemes, documents, organizations, personnel, drawings, predictions, estimations, rules, raw materials, products, etc). Each activity can be broken down to sub activities, which describes the detail of that activity. So, diagram can take hierarchical structure as in Figure 10.2 in order to describe large system.

10.2.2 Disciplines of IDEF0

IDEF0 representation is simple and understandable. Once model is properly created, it enables effective communication among stakeholders in the process. Although IDEF0 requires the use of a very simple notation, the way that current use of this notation set remains problematic. Following are the disciplines of persistent model in IDEF0 for effective communication (Kumagai, et al., 1998).

(1) Focalization: Control of Viewpoint

The creation of a useful IDEF0 model requires a single viewpoint presiding over the entire process. A successful IDEF0 model results when clear terms of one and only one viewpoint obtain, since in any other terms the quality of the system description is dramatically reduced. Two activities in Figure 10.3 have the same kind of ICOM arrows,

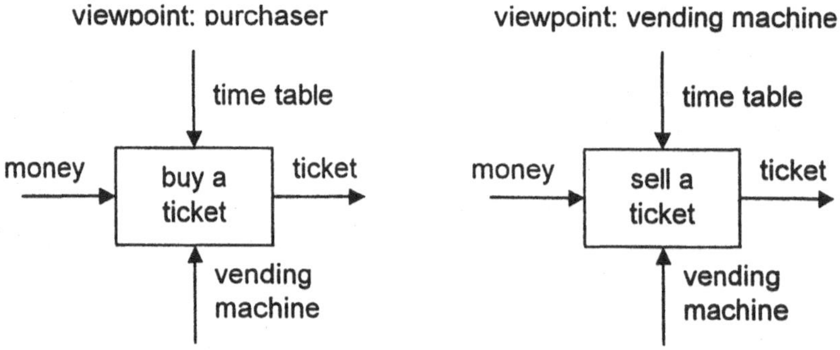

Figure 10.3 Difference of viewpoint

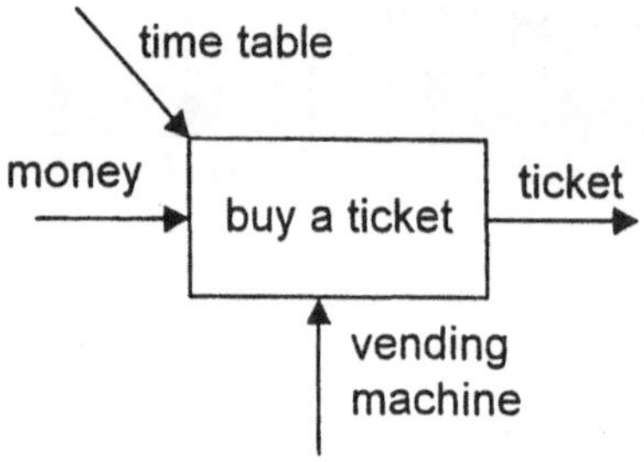

Figure 10.4 Is timetable located as input or control?

but different viewpoints lead them to completely different models. The left box would be broken down to user's activities of vending machine such as "count money", "feed money", and "push bottoms", etc. The right box would represents detailed activities of vending machnes, including "sense and count coins", "wait for user to push botton", "count coins", "output ticket", etc.

(2) Clear Classification of Material and Information
Generally, because any ICOM is subject to interpretation its meaning is intrinsically reader-dependent. Say that the reader views the activity in the example (Figure 10.3) in terms of the vending machine's viewpoint, "ticket" would be interpreted as its physical material because further decomposition would imply specification of the vending machine. If purchaser's viewpoint is taken, the reader would interpret "ticket" as information about the ticket.

(3) Clear Classification of Inputs and Controls
Because an ICOM is merely a label, how to classify an ICOM is often discussed by the modeling author and reviewers during modeling. Particularly ambiguous are the criteria that define whether an ICOM is control or input. In the example, the timetable may be considered unchangeable and therefore handled as a control. But if checking a timetable and selecting a train becomes dominant, the activity becomes posited as "to buy a ticket at a station," and so "timetable" can also be handled as an input (Figure 10.4). The input

case holds that the information content of "timetable" is converted. This ambiguity would revoke the issue (2).

10.3 TOOL FOR AUTOMATIC GENERATION OF IDEF0 DIAGRAMS

We have developed an experimental tool to automatically generate IDEF0 diagrams. The main target of our tool is a simple order entry system, which reads records from files and writes record to files using Web pages as user interface. The tool inputs the information about input and ouput of data items on Web pages, and generates an IDEF0 diagram.

10.3.1 Correspondence between data item and ICOM

A target system is assumed to satisfy the following statements:

- It mainly refers to and updates files based on Web page input and output,
- It peforms collective tasks on one Web page,
- Each data item is inputted from at most one file, and outputted to at most one file.

In our tool, one Web page corresponds to one activity of IDEF0, and arrows are determined based on the attributes of the data items in the Web page.

Focusing on each data item, three steps, display, input (selection), and save, exist. The following three cases can be considered about display step:

A1 Nothing is displayed. It assumes that a user inputs data from a keyboard.
A2 The data dependent on the internal system state or other data item is displayed.
A3 A certain file is referred to.

The following four cases can be considered about input (selection) step.

B1 Nothing is inputted. The displayed data is passed to the saving step as it is.
B2 A user inputs data.
B3 A user chooses from displayed data.
B4 Depending on the internal system state or other data items, data is chosen automatically.

About saving step, it is only the following two cases.

C1 A data item is only displayed on a screen and is not saved.
C2 A data item is saved in a file.

Table 10.1 covers all the combinations of the above three steps. There are three kinds of impossible combinations shown as †1, †2 and †3 in the table. These combinations are impossible because of the following reasons:

†1: Display A1 can only be combined with input B2, and vice versa, because display A1 requires user's input. On the other hand, display A2 and A3 do not need any user's input.
†2: The combination of display A3 and input B1 means full copy of a file, which is not the purpose of the target system.
†3: The combination of display A2 and input B4 means that the data is chosen automatically, which is equivalent to the combination of display A2 and input B1.

Except the above impossible combinations, there are five kinds of data items (mXX) with C1 when the result is only displayed and not saved, and five kinds of data items

(MXX) with C2 when the result is saved into a file. Each kind of data item is processed as follows:

m12, M12:
A user inputs data.
m21, M21:
Data displayed by the system is used as it is.
m23, M23:
A user selects data from those displayed by the system.
m33, M33:
A user selects data from those read from a file.
m34, M34:
The system automatically selects data from those read from a file.

Arrows of IDEF0 are made to correspond under the following rules to the above ten kinds of data items.

(1) Make a reference file (A3) correspond to a mechanism.
(1) Make a data input (B2) correspond to input.
(2) Make the selection by the user (B3) correspond to control.
(3) Make a screen display (C1) correspond to output.
(4) Make the saving to a file (C2) correspond to both a mechanism and output.

Arrows of the data item of mXX and MXX are shown in Figure 10.5.

Table 10.1 Classification of data item

		A1	A2	A3
B1	C1	†1	m21	†2
	C2	†1	M21	†2
B2	C1	m12	†1	†1
	C2	M12	†1	†1
B3	C1	†1	m23	m33
	C2	†1	M23	M33
B4	C1	†1	†3	m34
	C2	†1	†3	M34

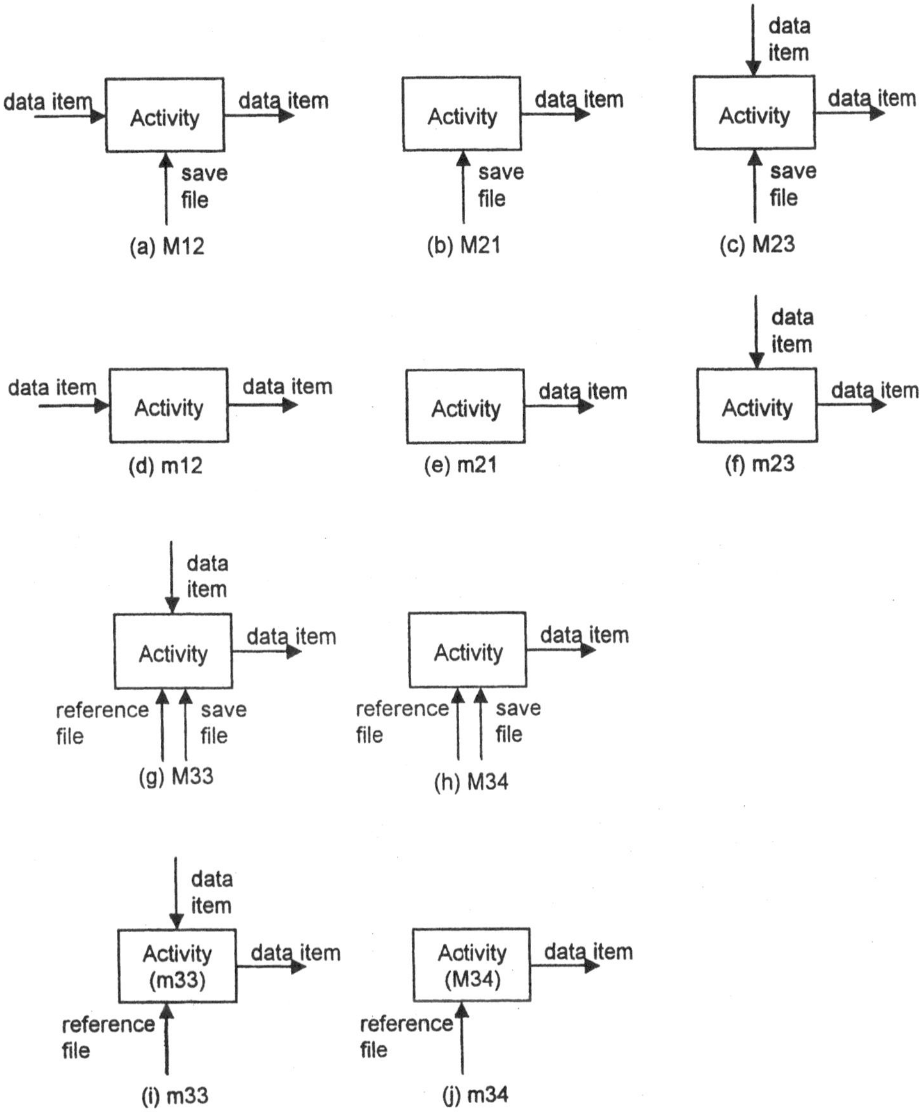

Figure 10.5 Correspondence between data items and ICOM

10.3.2 Connection of activities

When the output item name of a certain activity and the control item name of another activity are the same, we suppose that the former output is controlling the latter, and we connect the output arrow with the control arrow.

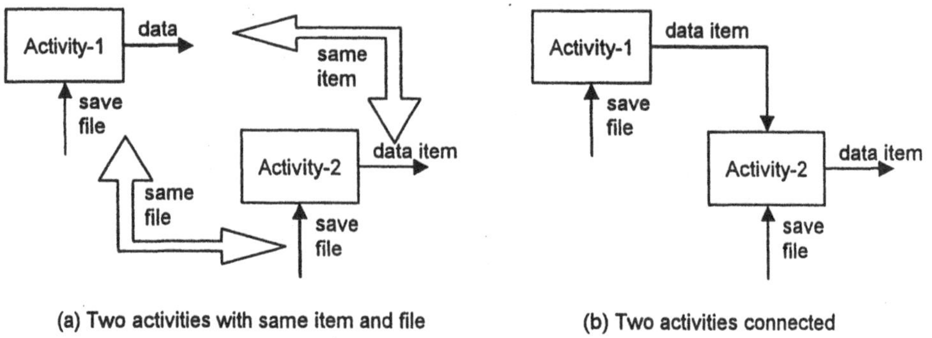

Figure 10.6 Connection of two activities

There is implicit connection through a file between activities. It is the case where a certain activity saves a data item to a file, and another activity refers to the file about the same data item. Even if the latter activity has no input arrow, we suppose that the data item is passed from the former activity to the latter activity through the file (Figure 10.6).

On the other hand, when the two activities share the same input, control, output, or mechanism, we do not put an arrow from one activity to the other activity because we cannot suppose any order of the two acitivities.

10.3.3 Tool for automatic generation of IDEF0 diagrams

We have developed an experimental tool for generating IDEF0 diagrams. A user has only to specify each data item of Web pages, and then the tool can generate an IDEF0 diagram. Figure 10.7 shows the sequence of input screens of the tool. Hereafter, a use case of the tool is explained according to the screens.

(a) Register a screen name of each Web page used by the target system. Each page corresponds to one activity.
(b) Register data items of each Web page. The name of each data item should be unique in the whole system
(c) Specify the kind of a display, an input (selection), and saving for each data item. The kind of data item is decided by this combination.
(d) Specify a reference file and a save file for the data item using files.
(e) Check the inputted data shown on the screen.
(f) The activities and the arrows which are generated based on input data are displayed. After the user certifies this generation, the tool performs automatic generation of a diagram succeedingly.

In addition, when two or more data items become the completely same arrow, that is, the arrows have the same starting point and the same terminal point, these arrows are gathered into one arrow by the tool.

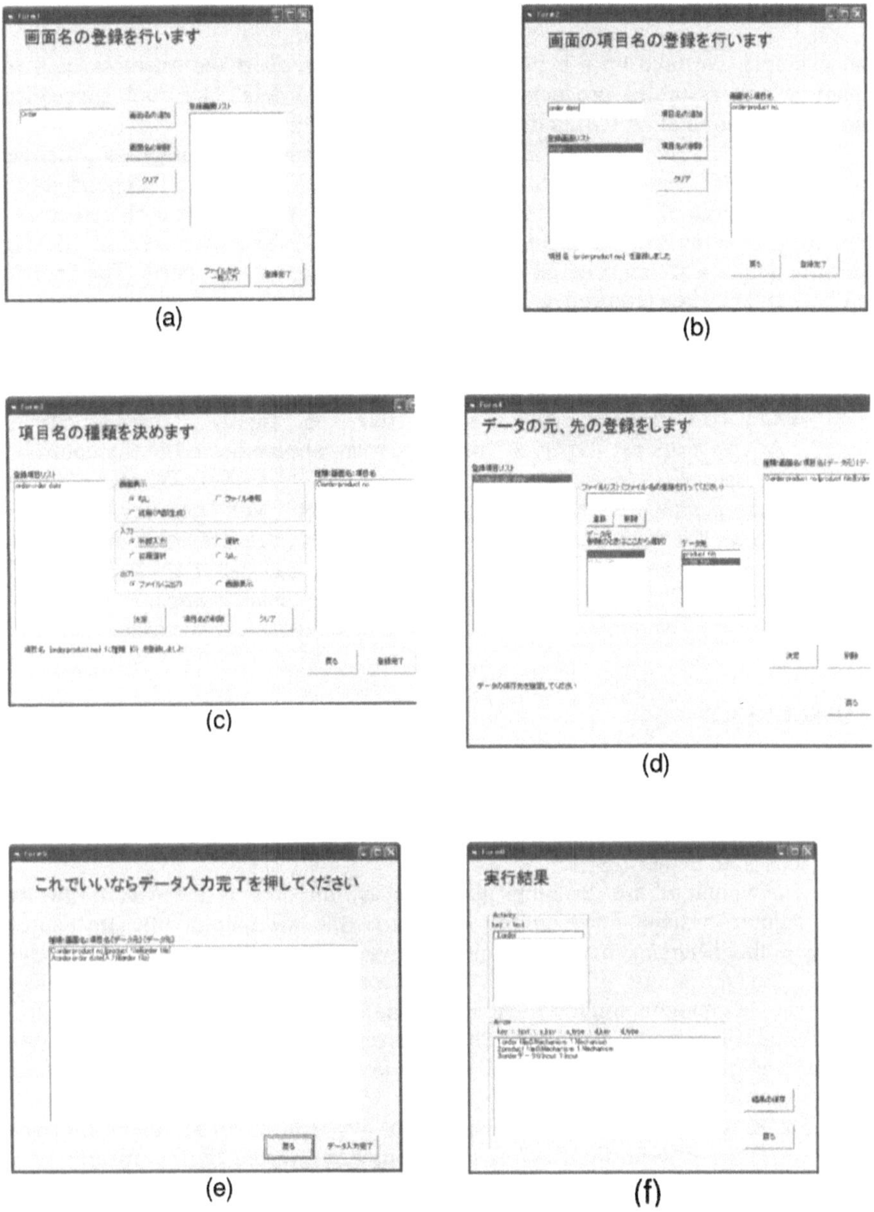

Figure 10.7 Input screens of our tool

10.3.4 Example of tool usage

As an example, we take up a hypothetical system to support the business of a small company which assmbles products according to the orders. Our tool succeeded in automatic generation of an IDEF0 diagram for the example system.

The example system has six screens, "order", "supply", "manufacture", "delivery", "cost", and "stock" (Figure 10.8). Suppose that the screen is working independently and the person in charge of each process inputs the item information on each screen in this system. When he finishes the input of "order", "supply", "manufacture", or "delivery", the information in a screen is saved into a file (it is the "order file" here). The "cost" and "stock" are called when required, and they display the cost of a product or the quantity of stock.

After specifying screen name, data item, and the kind of a data item, our tool generates an IDEF0 daigram shown in Figure 10.9.

In Figure 10.8, there are flows from "order" to "supply", "manufacture", and "delivery." Arrows from the output of "order" activity are connected to the controls and the output of "supply", "manufacture", and "delibery" activities. On the other hand, there is no mutual flow between "supply", "manufacture", and "delivery." This means that ther is no means to check the order relation in the three activities in the example system. Since what is necessary is to actually see a thing and just to do work, the lack of the order relation will not cause any problems if it is a small company. However, the viewpoint of business support shows that the system should check the order relation.

10.4 DISCUSSION

10.4.1 Disciplined IDEF0

From the viewpoint of the disciplined IDEF0 described in 10.2.2, we discuss the IDEF0 diagram generated by our tool.

The viewpoint of the diagrams generated by our tool is the user of the target business support system. For example, in Figure 10.8, we cannot find any confusion according to the difference in a viewpoint. However, if the scale of company or system is expanded, the viewpoints of the users will become different ones according to their positions, i.e., a common employee, an administrator, or a system administrator. In such a case, if the screens of the users who are in different positions are mixed, the diagram in which the viewpoints are intermingled will be generated. The measure for preventing such confusion will be needed.

In Figure 10.8, there is no confusion of a "physical material" and "information" because we focus only on the flow of the information treated by the business support system. However, since the generated diagram describes only the information flow, it may be an insufficient diagram in the viewpoint of grasping the whole business. As a further study, we will consider to generate the diagrams which represent the flow of "materials", and integrate them with the diagrams of this research.

In this research, we have treated the data passed through key input and files as the input of IDEF0 activity, and the data selected on the screen by the user as the control. Therefore, in Figure 10.8, the order ID becomes the control of "supply", "manufacture", "delibery" and the other data about the order, "order output", becomes the input of them.

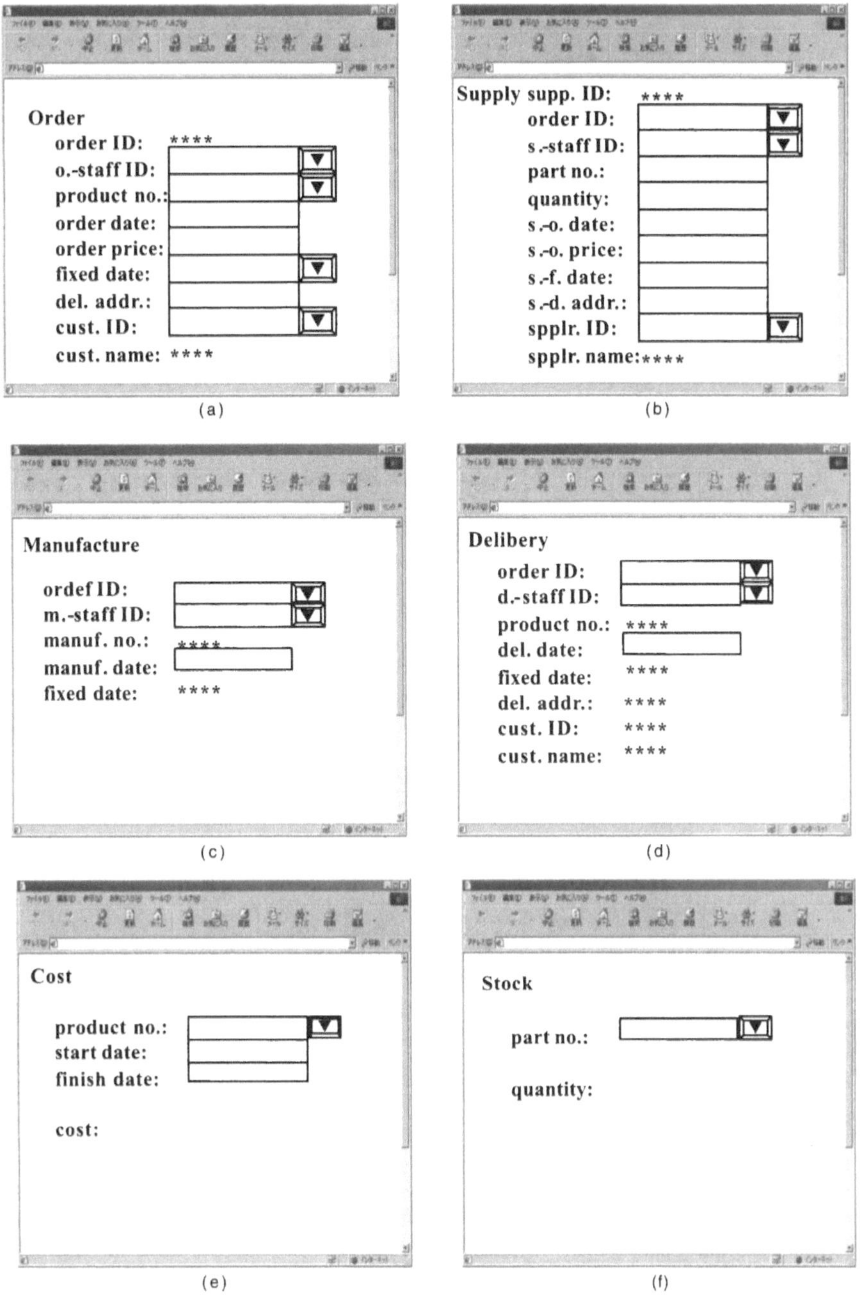

Figure 10. 8 Web pages of the example system

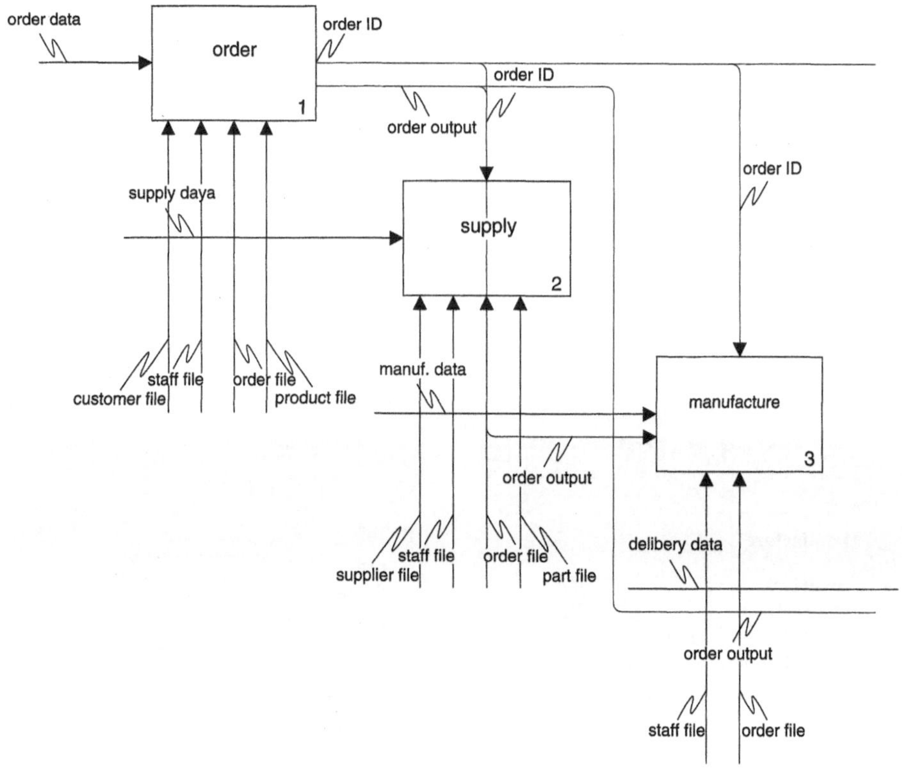

Figure 10. 9 Generated IDEF0 diagram

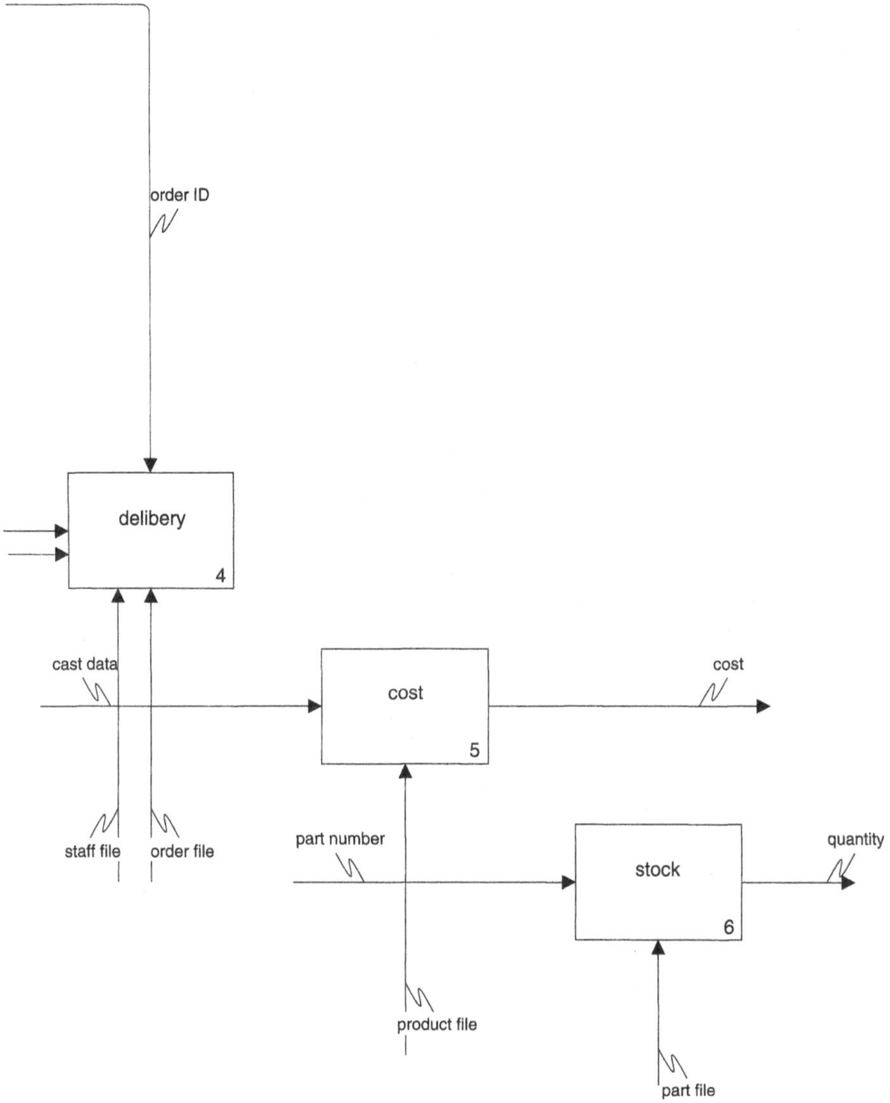

Figure 10.9 (continued)

In this case, the distinction of input and control agree with our common sense. However, to prove that this distinction is always appropriate, we should apply our tool to other types of systems.

10.4.2 Hierarchy of IDEF0

As our current tool cannot generate the diagram with multiple sheets, it can only treat a simple business. Hereafter we would like to introduce the hierarchy of IDEF0 described in 10.2.3. We suppose that the structure of a business organization will become a clue to divide an activity in the hierarchy.

10.4.3 Reuse of business model

As business models are generally more stable than those of information systems, they have an advantage in their reuse. In the case of the example described in 10.3.4, complete reconstruction of the system will be required if the amount of processings increases or the business is expanded. On the other hand, the business model which is shown as an IDEF0 diagram will be invariant, and we will just need to add a new element with the expanded business. From now on, it is necessary to examine this reusability concretely.

10.5 CONCLUSION

Based on the view that the business model is reflected in the existing business support system, we have devised the procedure which generates IDEF0 diagrams automatically, and developed a prototype tool.

We have analyzed data items in Web pages in the sequence of three phases, display, input, and save, and found 10 patterns of arrows (ICOM) for each kind of data items. Moreover, we have establishd the combination rules between activities. The first rule is to combine an explicit flow from an output to a control when both of arrows have the same name. The second rule is to combine an implicit flow through a file when one activity saves a data item into the file and the other activity read the data item from the file.

From now on, we would like to verify the usefulness of the generated diagram, and extend the procedure and the tool for corresponding to more complicated business.

ACKNOWLDGEMENTS

The authors wish to thank Mr. Isao Yamada of Yamatake Corporation for his helpful discussion, and Tomoyuki Tsuji who is a student at Kyushu Institute of Technology for helping us to develop the prototype tool.

BIBLIOGRAPHY

Itoh, K., Hirota, T., Kumagai, S. and Yoshida, H., 1998, Domain Oriented Systems Development: Principles and Approaches, in *the series of Advanced Information Processing Technology*, Vol.1 (Gordon and Breach Science Publishers).

Kumagai, Satoshi and Itoh, Kiyoshi. 1998, Designing Collaborative Work in IDEF0 using Interface Model, *Journal of Concurrent Engineering*, Vol.6, No.4, pp.333–343.

Marca, David A. and McGowan, Clement L., 1988, IDEF0/SADT Business Process and Enterprise Modeling (Eclectic Solutions)

CHAPTER 11

Generation and Conversion of Architectural Product Models†

Takeshi Ito

Kyushu Institute of Technology

Toyohiko Hirota

Kyushu Sangyo University

Masa-aki Hashimoto

Kyushu Institute of Technology

ABSTRACT

This study aims to build an environment which supports model conversion between data of CAD drawings and data of object-oriented CAD tools. Object-oriented CAD tools generally use object models as their internal models, but semantic and structural differences exist among them because each tool has its own purpose. In order to achieve model conversion among these kinds of data, we should construct a general purpose product model in which each part such as a column has its own attributes and the connections among the parts are described, and convert our model to that of a target CAD tool.

We have developed a system which supports model conversion between different CAD tools. We experimented with application of the generated object model in a system for designing a fire alarm system, which is a kind of object-oriented CAD.

11.1 INTRODUCTION

Various studies to recognize elements from drawings have been reported in the fields of image recognition and drawing interpretation (Nomura, et al., 1988), (Shio, et al., 1999), (Satoh, et al., 1992), (Wu, et al., 1996). Generating architectural product models is a kind of these studies. Knowledge regarding the size or length of elements, knowledge regarding the arrangement of elements, and knowledge regarding how

---

†This paper is modified from the paper presented at the SDPS conference on Integrated Design Process Technology, IDPT-2002.

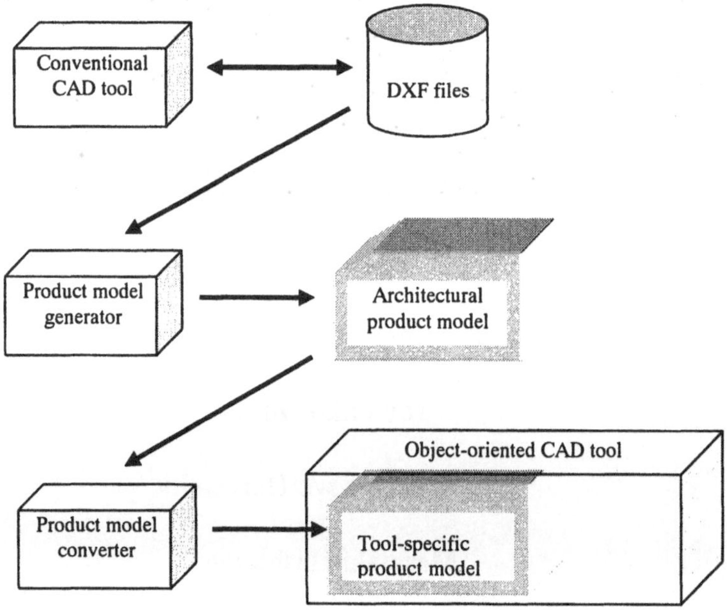

Figure 11.1 Data flow of our experiment

elements are drawn using basic figures such as lines, are necessary in order to extract drawing elements. However, the previous studies have no frameworks to describe each type of knowledge. All knowledge must be converted to the constraints on drawing elements in a describer's brain. Furthermore, he must consider the order of extracting elements. Against these disadvantages, we propose a declarative framework of knowledge description which separates the knowledge regarding a part itself, the knowledge regarding part connections, and the knowledge regarding the drawn figure.

We have developed a product model generator using these kinds of knowledge. Moreover, we experimented with application of the generated product model in a system for designing a fire alarm system. Figure 11.1 shows the data flow of our experiment.

In section 11.2, we explain the analysis of the architectural CAD domain. In section 11.3, we explain the framework of knowledge description. In section 11.4, we explain the method of generating product models. In section 11.5, we describe an experiment to prove the usefulness of our model. In section 11.6, we conclude our study.

11.2 ANALYSIS OF ARCHITECTURAL CAD DOMAIN

11.2.1 Domain analysis and modeling

In the field of engineering design - including mechanical design and building design - the designers are not supported effectively by a computer. There are many tools for drafting a plan or calculating structural strength, but there are few tools to directly support design itself as a designer-directed, intelligent activity.

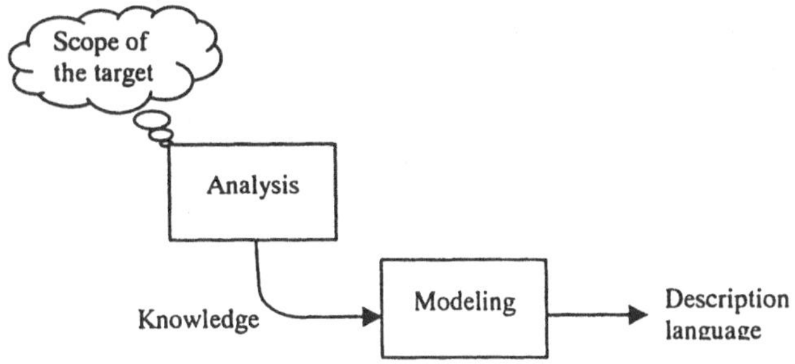

Figure 11.2 Domain analysis and modeling

As designer does not formalize his own design knowledge, developer cannot get sufficient knowledge from the designer interview for requirements analysis. A tool without design knowledge cannot be effectively used by the designer. Though useful tools exist for specific design tasks, and design knowledge is embedded in these tools, other persons cannot understand this embedded knowledge, and therefore cannot improve it.

To solve these problems and support intelligent design tasks, we analyze the design domain and construct a model on which domain knowledge can be properly represented. The developer's aim is to create a description language or a CAD generator, and the designer's aim is to put his design knowledge in order and describe it. If we develop a CAD system based only on the developer's knowledge, it may cause several negative effects such as a) a lack of knowledge about designer's practice, or b) embedding the design knowledge into the tool. On the other hand, if based only on the designer's knowledge, the resultant system may fall into failure because c) it is difficult for a designer to use general-purpose tools or languages, or d) analysis and modeling of the knowledge is improper.

To avoid the above situations, the developer and the designer must carry out their own domain analysis and modelling. However, if they are independent from each other, they cannot avoid the above-mentioned counter-product. Therefore they must cooperate with each other in order to carry out the process (Figure 11.2) (Itoh, et al., 1998). In the preparatory phase, analyst clarifies the scope of the target domain. In this case, it is the structural design of a medium- or lower-height building made of reinforced concrete.

In the analysis phase, he analyzes the target domain, investigates available technologies, and analyzes the designer's knowledge. As the output of this phase, he collects general knowledge. For example, a building is designed based on the reference lines, and a building is constructed by assembling parts, and the parts are dependent on each other.

Then, in the modelling phase, he specifies the language to describe attributes of each part and the relations between parts. This description language is not only the input to the subsequent phase, but also the input to the analysis for designer. The developer will develop a CAD generator based on the description language. While the designer describes his design knowledge by using the description language specified, the CAD generator automatically generates a building design CAD from the designer's described knowledge.

```
PART EastWestBeam
PLUG
  eastside_column into Column::westside_beam;
          ..............................
SOCKET
          ...        take        ...
END
ATTRIBETE
  width := default(300);
  east_x := eastside_column->west_x;
  west_x := ...
  height := default(600);
  north_y := (eastside_column->north_y
            + eastside_column->south_y) / 2 + width / 2;

          ..............................
END
DISPLAY
  polygon([{west_x, north_y}, {east_x, north_y},
           {east_x, south_y}, {west_x, south_y}]);
          ..............................
END
ENDPART
```

Figure 11.3 Example of BDL description

11.2.2 Knowledge regarding the building structural design

We have studied domain-specific specification languages in order to develop applications efficiently in the same domain, and we have already defined BDL (Building Design Language) (Hashimoto, et al., 1996) in order to develop the CAD tools which support architectural structural design.

In an average building, parts are located on a reference line in each direction X and Y at each floor. As the design process proceeds, other parts are subsequently located relative to a previously located part. In other words, buildings are constructed using a combination of parts and the connection relationships between the parts are specified. An example of BDL description is shown in Figure 11.3.

The connection relationships are described with PLUG-SOCKET in BDL descriptions. For example, the existence of a beam depends on the columns supporting the beam. Experts on structural design consider such existence dependency. A structural part on which the existence of other parts depends has a SOCKET statement in the relationship descriptions. On the other hand, a part whose existence depends on other parts has a PLUG statement. By using the PLUG-SOCKET relationship, not only connections between parts but also existence dependency and attribute dependency can be described. In the following, we explain the attribute dependency in detail.

Attributes of parts are classified into the following two groups:

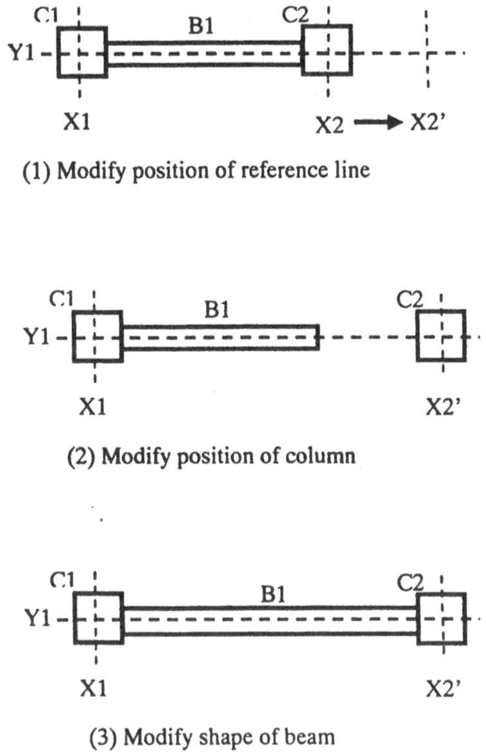

Figure 11.4 Example of attribute dependency

- Basic attribute: A designer determines its value.
- Derived attribute: Its value is obtained from basic attributes and other derived attributes whose values are already decided.

In Figure 11.4, when the position attribute of reference line X2 is modified, the position attribute of column C2 and that of beam B1 have to be re-calculated. These attribute dependencies are described with expressions.

To draw a part, attributes of the part are assigned to the parameters of drawing functions such as 'polygon' in Figure 11.3.

Thus, BDL has a specialized model of ER (Entity-Relationship) model (Chen, 1976); PART is a specialization of Entity, and PLUG-SOCKET is a specialization of Relationship. Furthermore, attribute dependency can be described directly. Various object-oriented CAD tools for structural design are efficiently developed with the descriptions in BDL and the operational descriptions for creating and deleting parts.

11.2.3 Association between buildings and drawings

We analyzed the building itself expressed with an average drawing in Figure 11.5. We listed 'x reference line', 'y reference line', 'column', 'east-west wall', 'north-south wall',

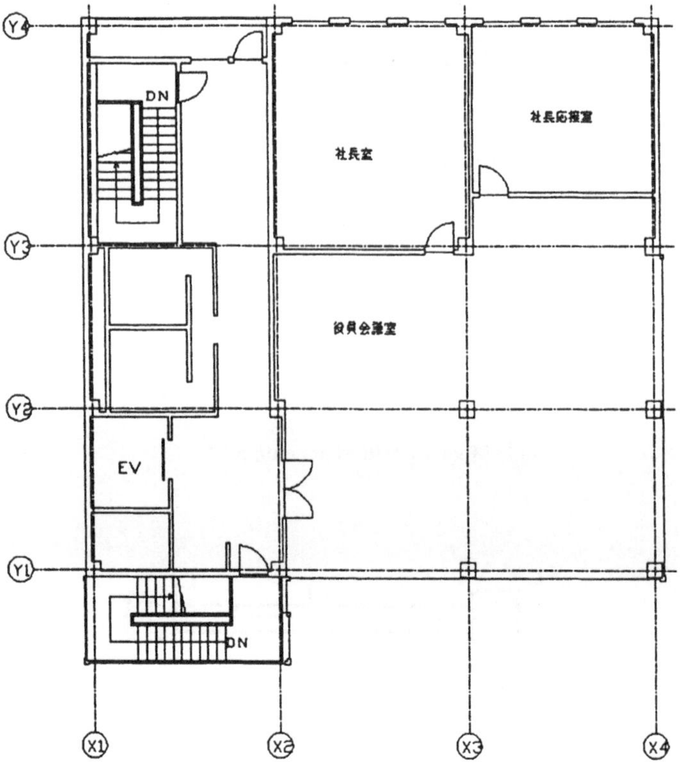

Figure 11.5 Sample CAD drawing for analysis

'east-west door', 'north-south door', 'east-west window', 'north-south window', and 'wall connector', as classes. The class diagram is partly shown in Figure 11.6.

A solid drawing in Figure 11.7 shows several relationships between parts; a column and north-south wall are connected with an east-west wall, and the east-west wall includes an east-west door and an east-west window. The drawing also shows constraints between attributes of the parts; the faces of the east side of the column and the west side of an east-west wall correspond, the width of the column is greater than the east-west wall, and the width of the east-west wall corresponds to those of the east-west window and the east-west door.

When designers make a drawing with a drawing tool, to express parts, they draw figures constructed from lines and arcs. Figure 11.8 is an expression of Figure 11.7 in a CAD drawing.

In CAD drawings, sections of parts one meter above the floor are projected, and the lines of interference positions between figures are erased. Only separated lines then remain and are saved in a CAD file. For example, the column with thick lines in Figure 11.8 is expressed with 5 lines.

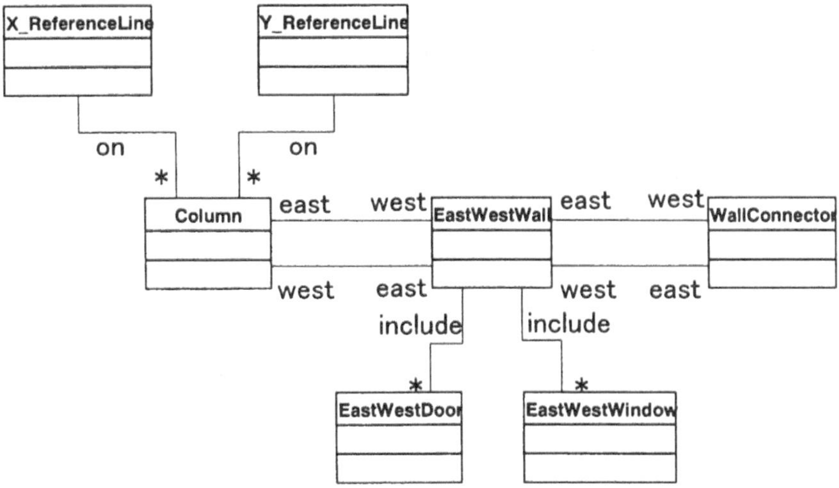

Figure 11.6 Class diagram of parts

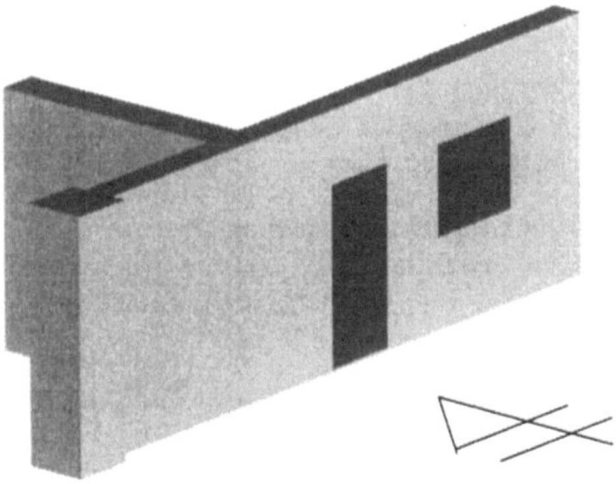

Figure 11.7 Sample solid drawing

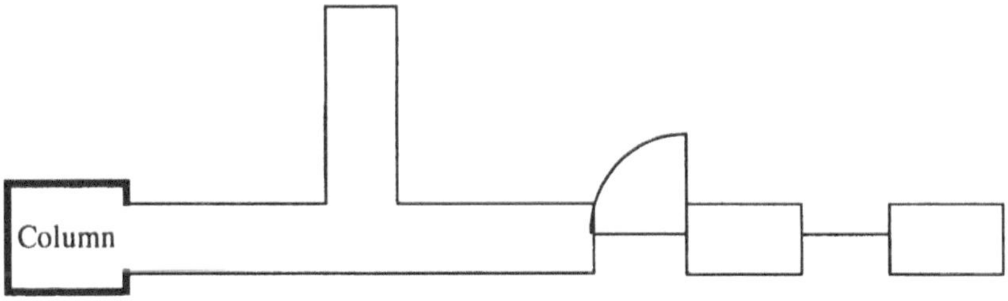

Figure 11.8 Drawing of Figure 11.7

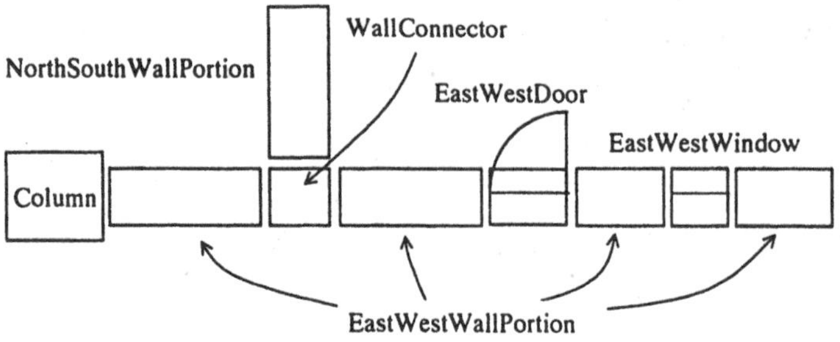

Figure 11.9 Division of walls

11.3 GENERATION OF ARCHITECTURAL PRODUCT MODELS

11.3.1 Knowledge for generating architectural product models

We classified the knowledge for generating product models into the following three types:

(1) Knowledge concerning each part
This knowledge describes constraints concerning the size of each part. The constraints are expressed in terms of the part's attributes.

(2) Knowledge concerning connections between parts
This knowledge describes constraints between the architectural parts that are connected. The constraints are expressed in terms of the attributes of the connected parts.

(3) Knowledge concerning drawn figures
This knowledge describes how the part is drawn using lines, arcs, and so on.

Because a wall including doors and windows as shown in Figure 11.6 is projected on a level face, its drawing is complex as shown in Figure 11.7. We divide a wall into wall parts such as 'east-west wall portion' or 'north-south wall portion' as shown in Figure 11.9. By dividing walls, all the relationships between parts become connections. The class diagram shown in Figure 11.6 is modified into Figure 11.10.

11.3.2 Knowledge descriptions

We show an example of our knowledge descriptions which have three frames corresponding to knowledge types

(1) Part knowledge frame
Attributes of the part are declared and constraints of the attributes and the range of each attribute value are described. Figure 11.11 shows the knowledge description of an east-west wall portion. The width of the part can take between 200 mm and 600 mm, and the function 'EastWestParallel' which is defined in the knowledge concerning drawn figures is called.

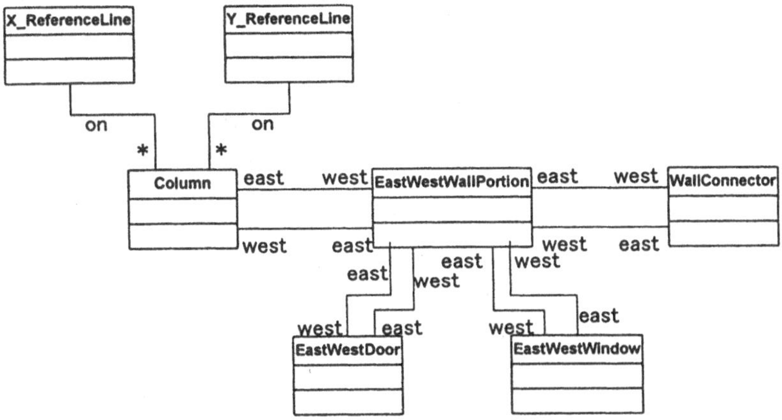

Figure 11.10 Modified class diagram to generate product model

```
part_knowledge('EastWestWallPortion',
        attributes([
         attr('east_x', EastX),
         attr('west_x', WestX),
         attr('north_y', NorthY),
         attr('south_y', SouthY)
        ]),
        constraints([
         WestX < EastX,
         WidthY = NorthY – SouthY,
         200 =< WidthY,
         WidthY =< 600
        ]),
        display('EastWestParallel', [
         attr('east_x', EastX),
         attr('west_x', WestX),
         attr('north_y', NorthY),
         attr('south_y', SouthY)
        ])
        ).
```

Figure 11.11 Example of part knowledge frame

(2) Part connection knowledge frame
Constraints between attributes of the connected parts are described. Figure 11.12 shows the knowledge description concerning the connection of a column and an east-west wall portion. 'east_x' of a column is equal to 'west_x' of the connected east-west wall portion, and the width of the column is greater than the east west wall portion.

```
part_connection_knowledge(
        ‘Column’, ‘westside’,
         attributes([
          attr(‘east_x’, ColumnEastX),
          attr(‘north_y’, ColumnNorthY),
          attr(‘south_y’, ColumnSouthY)
         ]),
        ‘EastWestWallPortion’, ‘eastside’,
         attributes([
          attr(‘west_x’, EWWallPortionWestX,
          attr(‘north_y’, EWWallPortionNorthY,
          attr(‘south_y’, EWWallPortionSouthY
         ]),
        constraints([
         ColumnEastX = EWWallPortionWestX,
         EWWallPortionNorthY =< ColumnNorthY,
         ColumnSouthY =< EWWallPortionSouthY
        ])
        ).
```

Figure 11.12 Example of part connection knowledge frame

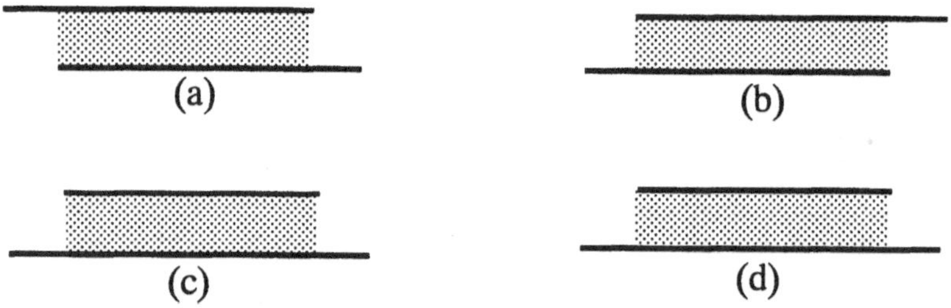

Shadow indicates an east-weat wall portion, but it is not drawn in CAD drawings.

Figure 11.13 Example of shape of east-west wall portion

(3) Drawn figure knowledge frame
East-west wall portion are drawn as shown in Figure 11.13. The knowledge description in Figure 11.14 concerns only (a) in Figure 11.13. 'Exist_object' is the predicate that asserts the existence of a basic figure, and 'ref_attValue' refers to an attribute of the basic figure.

```
drawn_figure_knowledge('EastWestParallel',
        attributes([
         attr('east_x', EastX),
         attr('west_x', WestX),
         attr('north_y', NorthY),
         attr('south_y', SouthY)
        ]),
        constraints([
         exist_object('east_west_line', NorthEWLine),
         exist_object('east_west_line', SouthEWLine),
         ref_attValue(NorthEWLine, 'y', NorthY),
         ref_attValue(NorthEWLine, 'east_x', EastX),
         ref_attValue(NorthEWLine, 'west_x', NorthLineWestX),
         ref_attValue(SouthEWLine, 'y', SouthY),
         ref_attValue(SouthEWLine, 'west_x', WestX),
         SouthY < NorthY,
         NorthLineWestX =< WestX
        ]),
        ).
```

Figure 11.14 Example of drawn figure knowledge frame

11.4 PRODUCT MODEL GENERATION

Our product model generator generates architectural product models referring to the above-mentioned knowledge. Its outline is shown in Figure 11.15.

11.4.1 Extraction of shape data from DXF files

DXF (Drawing eXchange Format) is the CAD drawing format specified by Autodesk Co. DXF is public and is supported by other CAD tools; exchanging DXF files instead of paper drawings is currently very popular. DXF is a kind of vector image format which includes coordinates of lines, arcs, and so on. In this study, we extract only necessary data to generate product models from DXF files.

11.4.2 Generation of architectural product models

In our method, it is important to select the object from which the reasoning starts. We select a column on intersections of X and Y reference lines. Our method consists of the following steps:

(1) At any intersection of X and Y reference lines, a column object is generated if the constraints of the part knowledge and the drawn figure knowledge descriptions are satisfied. The generated objects constitute the initial product model and are also added to the set N which is initially empty.

(2) One object P is brought from the set N.

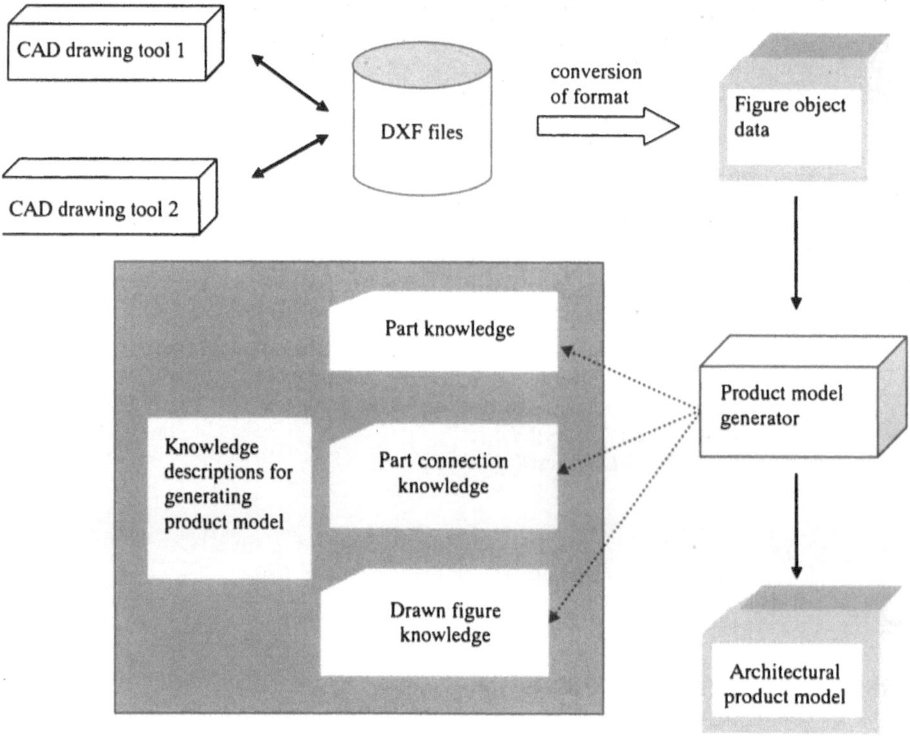

Figure 11.15 Flow of product model generation

(3) The classes C1, C2, ... Cn are selected which can be connected with P, referring to the part connection knowledge description of the class Cp to which P belongs.

(4) When the link with the same role as class Ci has already been generated, Ci is removed from the selected classes.

(5) When there is no link for the class Ci,
(a) an attempt is made to generate a part object which satisfies the constraints of the part knowledge and drawn figure knowledge description of class Ci, or
(b) an attempt is made to find an existent part instance which satisfies the constraints.

(6) If a new object is generated in the step (5), the object is added to the product model and also to the set N. Then, a link is generated between P and either the new object or the existent object, and it is added to the product model. If step concerning (5) fails, step is returned to (3) and the process is executed following step (4) concerning Ci+1.

(7) Steps (4) to (6) are repeated until all the classes selected in step (3) are processed.

(8) Another object P is brought from the set N, and step is returned to (3). If the set N becomes empty, this procedure is completed.

After a column object is brought from set N, an east-west wall portion object and a link between the column and the wall portion are generated. The final result is illustrated in Figure 11.16, and its internal form, i.e., Prolog facts, is partly shown in Figure 11.17.

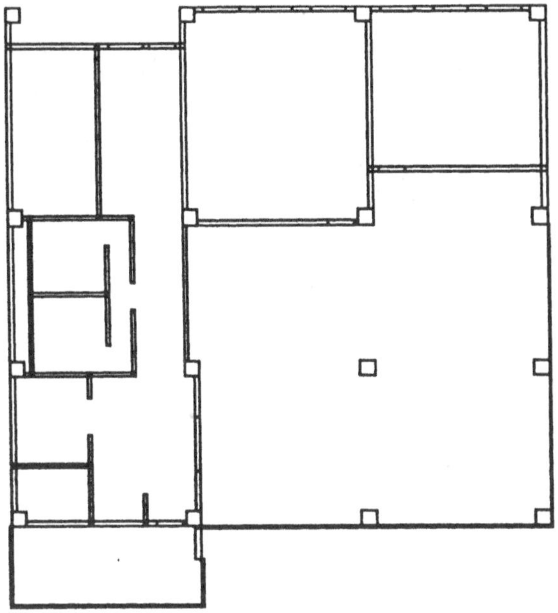

Figure 11.16 Result of product model generation

%architectural objects
object(1, 'WallConnectionPortion', [27660.0,27540.0,29160.0,29040.0]).
object(2, 'EastWestWallPortion', [29325.0,27660.0,29160.0,29040.0]).
object(3, 'Column', [29475.0,29325.0,29175.0,29025.0]).
object(4, 'EastWestWallPortion', [29525.0,29475.0,29160.0,29040.0]).
object(5, 'NorthSouthWallPortion', [29460.0,29340.0,29025.0,26775.0]).
•
•

%links
object_link(1, 'WallConnector', 'westside', 2, 'EastWestWallPortion', 'eastside').
object_link(2, 'EastWestWallPortion', 'westside', 3, 'Column', 'eastside').
object_link(3, 'Column', 'westside', 4, 'NorthSouthWallPortion', 'eastside').
object_link(3, 'Column', 'northside', 5, 'NorthSouthWallPortion', 'southside').
•
•

Figure 11.17 Example of product model

11.5 PRODUCT MODEL CONVERSION

11.5.1 Target CAD tool

Building design is generally classified into four fields: architectural design, structural design, equipment installation design, and estimation. IBDS (Integrated Building Design System), supports these phases of building design. The object-oriented structural designing tools developed using BDL is a subset of IBDS. In addition, IBDS includes a spacial-arrangement designing tool in the fields of architectural design and a fire-alarm designing tool. They are developed using a modified BDL whose domain specific constraints are loosened. We have selected the fire-alarm designing tool as a target CAD tool. Using this tool, a user arranges wall objects and constructs rooms, and the tool automatically arranges fire alarm objects and connects them with wires via the optimal routes.

The fire-alarm designing tool requires wall objects which can construct closed circuit which correspond to rooms as shown in the right side of Figure 11.19. These wall objects are different from walls in our product models. To distinguish the former from the latter, we name the former walls 'FADS-east-west wall' and 'FADS-north-south wall'.

11.5.2 Model conversion

We have developed a system which supports model conversion between BDL and IFC (Industry Foundation Classes) (Nakashima, et al., 2000). The latter is standard data specifications which can be shared with various architectural fields. The model conversion method of this study has adopted an approach similar to BDL and IFC conversion. To create object instances of the destination model from those of the source model, the schemata of both models must be described using the same schema description language. As the model of fire-alarm designing tool is described with BDL, we described the class definitions of our product model using BDL in this experiment. We also described rules to convert our product model to that of the fire-alarm designing tool using the operation language specified for IBDS. Moreover, we developed a format conversion program which converts our product model file into the IBDS save file. Based on these descriptions, our IBDS generator can generate a model conversion. Figure 11.18 shows the flow of conversion. The conversion itself consists of the following two phases:

(1) Conversion into east-west and north-south walls
Links between wall portion, windows and doors are already obtained. Therefore, a complete wall object is easily generated following the links.

(2) Conversion into FADS-east-west and FADS-north-south walls
The conversion of the walls obtained in (2) into FADS walls is shown in Figure 11.19. The following are the steps for the conversion into FADS-east-west walls, where the conversion into FADS-north-south walls has similar steps.

(i) One east-wet wall is selected.

(ii) Concerning the objects to which the selected east-west wall is connected,

(a) When the object is a wall connector, the end of the FADS-east-west wall becomes the center of the wall connector.

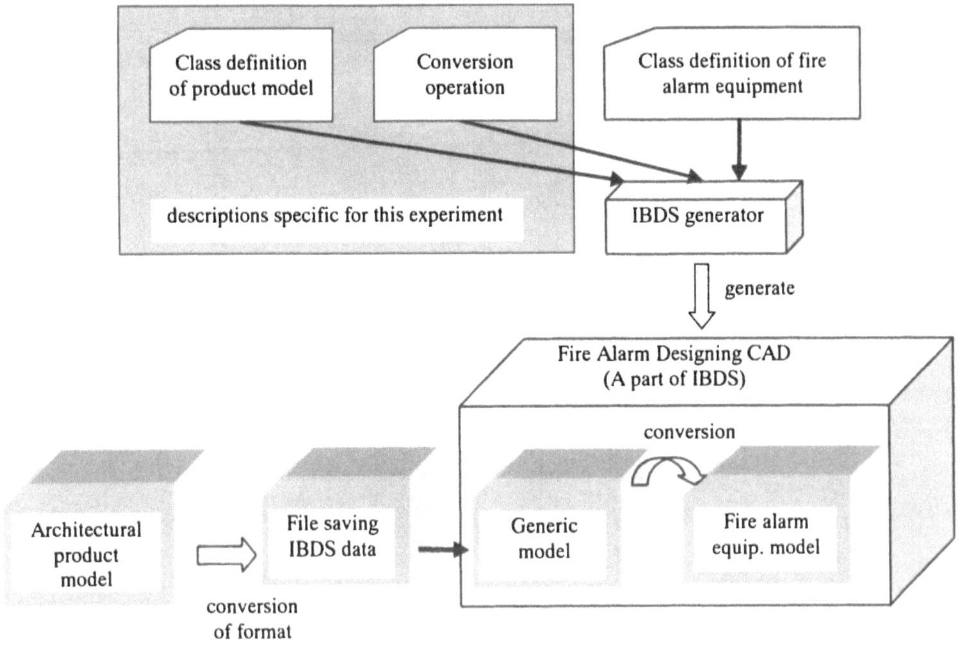

Figure 11.18 Flow of model conversion

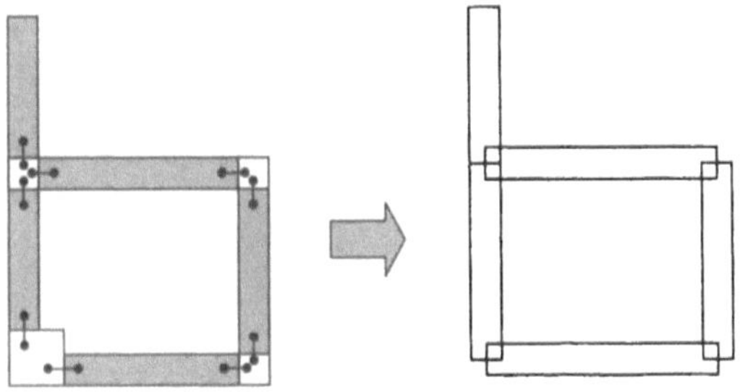

Figure 11.19 Extension of walls

(b) When the object is a column and is connected to a north-south wall, the x-coordinate of the end of the FADS-east-west wall becomes that of the center line of the north-south wall, and the y-coordinate becomes that of the east-west wall.

The results of the model conversion and the automatic designing of a fire alarm system are shown in Figure 11.20. The circles drawn at the center of the rooms are fire sensors. The receiver at the right upper side is connected to the two sensors by an optimal route. It is demonstrated that fire alarm equipment is automatically designed based on the wall objects of our product model obtained from a CAD drawing.

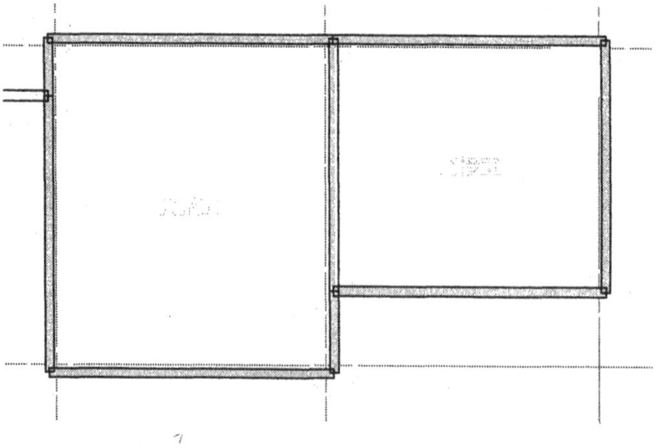

Figure 11.20 Result of model conversion and automatic design

11.6 CONCLUSION

We have proposed a method of generating product models for the support of data conversion between architectural CAD drawings and object-oriented CAD tools. Our model conversion can be applied to various phases of building design. For example, it will reduce the cost of the estimation phase by automatically counting parts in a drawing. In this paper, we have shown an application of this method to fire alarm system design.

Though there remains the problem regarding the knowledge description of drawn figures, a user has only to describe the knowledge concerning each part, the connections between parts, and drawn figures declaratively, i.e., he does not need to consider the procedural sequence for generation.

BIBLIOGRAPHY

Chen, P. P., 1976, The Entity-Relationship Model: Toward a Unified View of Data, In *ACM Trans. Database Systems*, Vol.1, No.1, pp.9–36.

Hashimoto, M., Hirota, T, et al., 1996, A Domain-Specific Conceptual Model In Building Design, *Proceedings of the Ninth International Conference on Industrial and Engineering Applications of Artificial Intelligence and Expert Systems*, pp.71-76.

Itoh, K., Hirota, T., Kumagai, S. and Yoshida, H., 1998, Domain Oriented Systems Development: Principles and Approaches, in *the series of Advanced Information Processing Technology*, Vol.1 (Gordon and Breach Science Publishers).

Nakashima, T., Efendy, L., et al., 2000, An Experience Report on an Implementation of Architecture CALS. In *Proceedings of the Fifth World Conference on Integrated Design and Process Technology*, pp129-136.

Nomura, M. and Tsukio, Y., 1988, Understanding architectural drawings using knowledge. In *J. of Japanese Society for Artificial Intelligence*, 3(4):494-502 (in Japanese).

Satoh, S., Ohsawa, Y. and Sakauchi, M., 1992, A roposal of Drawing Image Understanding System Applicable to Various Target Drawings. In *Trans. IPS*, Japan, 33(9):1092-1102 (in Japanese).

Shio, A. and Aoki, Y., 1999, Sketch Plan: A Prototype System for Interpreting, Hand-Sketched Floor-Plans. In *Trans. IEICE. Japan*, J82-D- II (3):431-439 (in Japanese).

Wu, W., Satoh, S. and Sakauchi, M., 1996, A Multipurposes Drawing Understanding System with Object-Oriented Model Description. In *Trans. IPS*, Japan, 37(6):1061-1071 (in Japanese).

Index